I0036503

International Project Management for Technical Professionals

Second Edition

By Brian E. Porter

ASME PRESS

Library of Congress Control Number: 2020941046

ISBN: 9780791883563

Dedication

To my wife who has helped keep the family running as I have traveled the world to fulfill my project responsibilities. To my three teenagers who have had to hear stories of places they have yet to see or experience. And to my friends in business from China, Tanzania, Romania and all parts of the globe! A special thanks to my friend and business associate Marcus Goncalves from Brazil who encouraged me to expand my horizons in teaching, consulting, writing and international business. Thanks to all who have helped me expand my experiences and therefore capability to share with others!

Brian E. Porter

Preface

This book has been intended to help other learn from my experiences as a project manager around the world. My travels have taken me to remote villages in Tanzania, major manufacturing in China and business hubs in the United Arab Emirates. The variety of experiences have shaped who I am and hopefully you will gain from them as well. In fact, one of my most frequent quotes is "smart people learn from their mistakes; wise people learn from others' mistakes". Hopefully you can learn from both the successes and failures noted.

This book is intended to be a practical tool rather than purely theoretical as one might find in many management books released today. Much of the book is based on practical experience, strategies and techniques attempted with various experience levels from interns and students to managing Ph.D. level engineers and scientists. Many of the experiences are special to me because of the situations that had to be overcome. As you read, you may identify situations that you connect with and say "yes, that has happened to me". International project management is bound to grow as the digital world permits more and more individuals onto the superhighway. As I'm editing this second edition, we're in the middle of quarantine for the COVID-19 situation. Recognizing the benefits of lower cost labor or improved productivity in another nation is old news. For centuries individuals have made their entire life substance by trading internationally via wagon, boat or train. Today the transportation mechanisms and communication tools are far more advanced, but the principles are the same. We can use digital media to advance the causes of project management.

For an individual in engineering, presumably in mechanical engineering for many interested in this book, the first piece of advice is to recognize that you will be challenged to balance engineering and project management. Step back from the product, equipment or service that you are in charge of and realize that it is the end-result of a project. Focusing on the project will help the product, but just as important, the time and budget delivery.

Chapter 1, Project Management, takes a look at the responsibilities and general characteristics of project management. Some individuals reading this book may only be interested in the international aspect, but we must take into account that some individuals will not have the background in formal project management and that the basics must first be illustrated. We also identify the broad certification categories and the benefits of PMI.

Chapter 2, Project Types, provides a look at how projects are similar and can be quite unique. The metrics of the broader project can open up the eyes of the narrower focus on the scope. If you take the broad view and realize the magnitude, complexity, domestic or international nature and other features, it will help you keep the big picture in mind.

Chapter 3, Cultural Bearings - Social, will delve into a travel channel type of approach that will address many of the differences that shock those who have not had significant exposure to other countries, even in a social capacity. For those that have traveled significantly, this will just be a reminder, but should at least bring to the surface some key issues to remember to ensure success.

Chapter 4, Cultural Bearings - Business, intends to expand on basic cultural adaptations that a project manager encounters. The settings in business require more attention and likely will enjoy less forgiveness of mistakes than for those in a social setting.

Chapter 5, Exploring Brazil, China, India and USA, takes a look at the cultural and statistical differences of operating in various countries. A project manager can look at the benefits and challenges of a country or multiple countries side by side. Even if a project manager is comfortable with the differences, he or she may wish to share such charts with the project team since many of the associates may be unprepared for the culture shock that may ensue.

Chapter 6, International PM Initiation and Planning Tools, begins with a look at the domestic side of project management and then shares further international techniques. Engineers who act as project managers tend to look at these tools as technical devices, but also need to recognize the balance between technical and human characteristics to effectively communicate the scope, schedule and budget plans from the first meeting through the end of the project.

Chapter 7, International PM Execution and Controlling Tools, provides an evaluation of variation execution and monitoring techniques that will be effective for both domestic and international efforts. The basics of domestic project management are shared since these skills should be mastered before entering international project management.

Chapter 8, Individuals Suited to IPM, delves into the aspects of engineers, project managers and the extra something that international project managers need to embody. While no project manager is perfect, one should strive to meet the primary requirements for the specific project at hand and make sure that the project team meets these basic principles.

Chapter 9, International PM Perspective shares a number of stories and lessons that have been experienced by the author and are shared as words of caution to help others learn without suffering the same consequences in their projects.

Chapter 10, International PM Resources, includes a list of additional websites, books, magazines and resources to tap in order to effectively improve the reader's project management skills starting immediately following the text. These resources are used by the author to initiate, plan, execute, control and close projects both domestically and specifically internationally.

Chapter 11, Good, Bad and Ugly Projects this is a new chapter that seemed relevant and will include a number of additional cases from around the world that a new or moderately experienced project manager should focus on to gain insights into how to run AND how not to run a project!

Chapter 12, Road Map for the "New PM" which focuses on a step-by-step recommendation on how to get involved in project management.

Who Should Read This Book

The intended audience is primarily technical staff such as mechanical engineers, chemical engineers, technicians, and technical managers that will be required to manage projects within the US and more specifically, international projects. The lessons taught within begin with a broad perspective for those that are new to project management and then look deeper into the specifics of international project management.

The beginner in project management may feel that some terms are not deeply explored. For those individuals, it is recommended to pursue an introductory course in project management which will explore the necessary tools and terms prior to seeking international project management. You may read this text and then look for a course many universities offer. ASME offers a course in various parts of the US several times per year and will organize an event for your company as well.

Those that are experienced in domestic project management will find the recommendations within this text practical. The intent is not to explore the theoretical aspects of project management. Rather, this is a Technical Survival Manual and shares the examples from real life that can be implemented immediately into projects.

Who Should Read This Book

Acknowledgement

This book was originally completed with the support of Marcus Goncalves, my partner and friend who has shown me what is possible by using a variety of skills, while focusing on the work at hand. He again has been a support and mentor to me.

Also, the experience of the masters program from Boston University has been beneficial. Professors, facilitators and students have helped to solidify a better understanding of mankind and how to work with multiple cultures and across virtual teams of all sizes. Over the past decade, I've continued with Boston University as a facilitator in a number of courses which have helped me to round out my skills.

A thank you and credit to the staff of ASME Press for their reviews. Appreciate the help making my "speak-writing" useable for readers!

Table of Contents

Chapter 1
Project Management Basics

You can have it good, fast, or cheap: pick any two. "The Project Manager's Maxim" [1]

In order to gain a grasp on project management, we must first have a basic understanding of what a project is and is not. Many individuals will claim that they work on projects, are a project manager or have too many projects to keep up. In many cases, they are correct about such a claim, but in other circumstances, there is simply a lack of understanding for what a true project is and the responsibilities associated with a project. There are a few criteria that must be identified early:

A project must have a defined start and finish. It is temporary. The first element that is frequently lost on individuals is that a project is not a part of operations and requires a distinctly defined start and finish to qualify as a project. All too often individuals will be a part of the operations function and they have several tasks to address every day, but when these tasks are part of an on-going, cyclical process, it is not a project. In fact, if a defined end-point is lacking, then it is not a project. To properly manage a project, there should be a well-defined schedule breaking down required efforts into smaller subsections and identifying intermediate deliverables. More on schedules later. [2]

A project must create a unique product, service or event. A project will eventually result in the desired work product, which can take many forms. Individuals who develop new products will eventually

1

see their concept turn into something that the end consumer can use. Professionals intend to create a new service that will provide for the needs of the consumer by easing a task or bringing something new or improved to the world. [3] Some project managers will organize an event such as a charity fundraiser. Each of these "work products" are involved in creating something new or different that has not been created before.

A project must be complex. This requirement is certainly subjective. Very few individuals would challenge the claim that planning, designing, construction and occupation of a 100-story skyscraper is "not complex" and would not meet the requirements of a project. Very few individuals would argue that tying your shoes each morning *is* complex or is a project. However, there are so many examples in between that one might be inclined to claim project status while others would be firmly set against such a claim. Experience in the particular industry would be crucial to this judgement call and ultimately those involved will have to make the determination on how to handle the specific assignment, as a project or as a process.

A project must be non-routine. The key word is "unique". If the product, service or event has occurred before and the planning will simply involve taking out the last schedule, adjusting the start date and following the steps almost identical to the first project, then there is no project involved. A process has been defined and the repetitive versions are variations on a project. Thus, those who work on a manufacturing line and perform the same inspections on a daily routine, will not qualify as managing a project. Both skills, project management and process management are absolutely necessary, but they are distinctly different roles and require different skill sets. [4]

A project will have a separate budget identified. In order to develop a new product, service or event that is new to the world, complex and non-routine will require a budget. This could be as small as a few hundred dollars for a non-profit organization that uses volunteer hours, but it could also be $42 billion as the Beijing Olympics demonstrated [5]. A defined budget should be prepared based on estimates from the project team, project manager and prior experiences. More on this later, as well.

The example frequently given to identify a project is that of changing a tire. If somebody changes your tire for you on the side of the road, they are providing you with a service. That may meet one of the definitions. You may never have changed a tire yourself, which may indicate that it is a "new" experience. You may have to pay for the tire to be changed and might indicate that there is a budget. However, we must consider one important element. Changing a tire is a "routine" topic simply based on the fact that there is an instruction manual provided, usually in the glove box, and thousands of tires are changed each day by individuals who follow the prescribed process to change the tire. One might also argue that this is not a project based on the fact that it is not too complex. The required steps are usually completed in a page or two. Thus, with a recipe, experience by millions of individuals worldwide and a budget that would consist of one line item, changing a tire is not a project.

NON-PROJECTS

	Start/ Finish	Unique	Complex	Routine	Budget
Changing a tire	Yes	No	Maybe	Yes	No
Making tires at a plant	No	No	Yes	Yes	Yes
Making a Denver Omelet	Yes	No	Maybe	Yes	Small!
Assembling a prefab bookcase	Yes	No	Maybe	Yes	Small!

MARGINAL PROJECTS

	Start/ Finish	Unique	Complex	Routine	Budget
Modest product improvement	Yes	Yes	Maybe	Maybe	Maybe
Addendum to existing work	Yes	Maybe	Maybe	Maybe	Yes

PROJECTS

	Start/ Finish	Unique	Complex	Routine	Budget
Construction of a skyscraper	Yes	Yes	Yes	No	Yes
R&D & Manufacture new toy	Yes	Yes	Yes	No	Yes
Organizing Olympic Games	Yes	Yes	Yes	No	Yes

Figure 1.1. Examples of projects and non-projects

Another example frequently referenced is the remake of a movie. Considering the gross lack of creativity in the movie industry today, producers frequently will fall back upon a past success to draw current success. Consider a movie that was made in the 1960s or 1970s that was incredibly popular. Would the remake of this movie be considered a project or not? Take a look again at the requirements:

A project must have a defined start and finish. Making the second movie would certainly require a start and finish schedule with the actors, crew and staff.

A project must create a unique product, service or event. This might be questionable, but anybody that has seen a remake knows that they rarely have the same quality as the original and therefore it is essentially a unique product. If it were not, why would individuals wish to see the original again?

A project must be complex. There should be little question that the making of a movie is a complex task. Numerous individuals, sets and resources must be well-coordinated for location, timeliness and quality of material.

A project must be non-routine. Gathering this particular set of actors, crew and staff is likely unusual. The location, sets and materials needed will be different than other movies made even if with the same basic actors and director. It is unlikely that the majority of other individuals are all involved, unless it is a sequel.

A project will have a separate budget identified. A movie absolutely requires a budget to cover all of the staff, materials, food, housing, etc. necessary to move a small city during the shoot.

Now that we have covered what a project is, we can begin to look a bit at the history of modern project management.

History of Project Management

Projects have been managed since the beginning of time, but there wasn't always a formal procedure of how to manage projects. It appears that over time there was a growing need to manage projects more formally as resources become more difficult to secure and control.

Consider some of the greatest projects of all times in the wonders of the world such as the Great Pyramids of Egypt and the Great Wall of China. The pyramids and the wall that separates China from Mongolia have to be considered some of the greatest feats of history simply because of the magnitude of the work required with primitive tools. However, they learned to use the tools available and used massive amounts of manual labor, which would not be economically possible in the twenty-first century.

During industrialization of the 19th century in the nations of Great Britain, Germany and the US, companies began to organize and implement upgrades of machinery and techniques into the massive production plants using rudimentary project management skills. [6] Let's take one example here. John Rockefeller required some serious project management skills to make the transcontinental railroad possible and with an attempt to remain efficient. Consider the sheer magnitude of moving a small city of surveyors, carpenters, welders, cooks and butchers that it took to keep up with the movement of laying a few miles of track every few days. The men would have to sleep in rolling carriages on the rails that were recently completed. There were herds of cattle that had to be moved with the men because the distance to move beef and other food-stocks would be too expensive. The cattle were even moved on the rail as it was being completed. Consider the details of communication for the westward moving crew with civil war veterans and Irish immigrants who needed to meet up with the eastward moving crew from San Francisco comprised heavily of Chinese immigrants. If either of the two teams of surveyors was not planning accurately, the eastward and westward rails could be off by several dozen miles. Instead, they had to be within fractions of an inch for everything to work properly.

Furthermore, consider that the railroad industry was being supported by the iron industry for the rails for Pennsylvania, lumber industry for the rail ties in Michigan and Wisconsin, boating to move the rail ties across Lake Michigan, cattle industry to feed the workers and numerous other industries. Planning, execution, controlling and communication were effectively implemented despite the lack of modern tools, by today's standards. Ultimately this effort changed a very dangerous three- month trip by wagon to one-week trip on a reasonably safe and less strenuous rail ride. The closing of this project was ceremoniously completed with the golden spike driven into the last rail, which was promptly removed and replaced with a standard iron spike for obvious reasons! This is one example that projects were existent and doing well. [7]

Looking forward to the early 1900s, many individuals marvel at the auto industry and Ford's production line. However, Ford is the master of process management and not project management. Production lines are necessary, *following a project*. Ford developed the Model-T and was happy with the results. In fact, the popular phrase from Ford's autobiography "Any customer can have a car painted any color that he wants so long as it is black" [8] carries on to this day. Henry Ford was not interested initially in creating numerous different cars and performance. His fame comes from finding ways to make that same vehicle more cost- effective and find ways to make sure that the product quality was identical from automobile to automobile. Hence, project management took a back seat at Ford during those years and process management or operations management was critical. There is plenty of evidence though that he moved back into projects and product development as well! Feel free to visit the Ford/Edison Museum if you ever visit Fort Myers.

Many other industrialization efforts such as buildings, railways, new factories and businesses required project management, which lead into process improvements during that period. My personal favorites are the inventions of Thomas Edison. The man was tireless and during the late 1800s and early 1900s he managed to secure over 1,000 patents. This is more than four times the number of patents issued to all inventors from 1790 to 1800, the first decade of the patent office in the United States! [9] Look at a list of products developed during that time by Thomas Edison (1,093 patents) alone. Consider the unimaginable list from Westinghouse (361 patents) [10] and many others [11]:

Edison's Inventions
✓ Automatic Telegraphy
✓ Cement
✓ Disc Phonograph
✓ Electric Generator
✓ Electric Lamp
✓ Electric Light and Power System
✓ Electric Pen
✓ Loud-Speaking Telephone
✓ Motion Pictures
✓ Ore Milling
✓ Quadruplex Telegraph
✓ Stock Ticker
✓ Storage Battery
✓ Telephone Transmitter
✓ Tinfoil Phonograph
✓ Vote Recorder
✓ Wax Cylinder Phonograph

Of course, similarly impressive lists of projects in construction, land development, business formulations and exploration could be presented for these periods, but that is not the intent of this book or chapter!

Consider the growth of computers and information technology during the 1950s. Look back to the examples that were made possible through the support of university research and private funding. Then, beyond simple computers run by punch card and operation, we see the internet conceptualize through the papers of 1965 by MIT's Joseph Carl Robnett Licklider and his outlines for the Galactic Networking Concept. [12] Each and every expansion into information technology, would allow individuals to expand their reach into unknown areas of the vast world-wide-web of data. These initial efforts lead to individual in the 1970s at IBM and Apple to develop the necessary hardware to support individual homes and locations. Then Microsoft and other companies in the 1980s focused on developing software platforms to support the user. Ultimately in 1994 the explosion of the internet allowed Joseph Licklider to see a dream come true. Individuals from around the planet with different computers, different systems, different monitors and different needs could all connect through this virtual world. In 2020, with the COVID-19 situation, many companies are able to continue to operate without meeting their customers. Training continues online for elementary to university

Today one can see project management in luxury construction projects throughout the US, information management through the national database and event management through projects such as the the Olympic Games.

As time passed, it became evident to many individuals throughout the United States and industrialized nations that formal processes were necessary to increase the efficiency under which projects would be managed. At first, individual organizations made their own efforts, but with time, society's developed to integrate and increase uniform terminology expanded throughout the US and Europe. Incepted in 1969 by working professionals, the Project Management Institute (PMI) has experienced 10-15% growth or more for four decades and continues to see expansion possibilities for years to come. [13] With 550,000+ members and growing, the organization recognizes eight specific, ISO-accredited designations [14]:

Certified Associates in Project Management (CAPM®)
- Understand the processes and terminology and have a fundamental knowledge of the *PMBOK® Guide.*
- Demonstrate knowledge of project management practices.
- Contribute to project team as a Subject Matter Expert.

Project Management Professionals (PMP®)
- Are responsible for all aspects of the project for the life of the project.
- Lead and direct cross-functional teams to deliver projects.
- Demonstrate sufficient knowledge and experience to apply a methodology to projects.

Program Management Professionals (PgMP) ®
- Are responsible for achieving an organizational objective by overseeing a program that consists of multiple projects.
- Define and initiate projects and assign project managers to manage cost, schedule and performance.
- Maintain alignment of program scope with strategic business objectives.

Portfolio Management Professionals (PfMP) ®
- Responsible for an organization's portolio
- Advanced experience for managing projects and portolios

PMI Risk Management Professional (PMI-RMP^{SM}):
A project risk management professional provides expertise in the specialized area of assessing and identifying project risks, along with plans to mitigate threats and capitalize on opportunities.
- Responsible for identifying project risks and preparing mitigation plans.
- Supports project management and the team as a contributing member.
- Minimum of three years of project risk management experience.

PMI Scheduling Professional (PMI-SP^{SM}): A project scheduling professional provides expertise in the specialized area of developing and maintaining the project schedule.
- Responsible for creating and maintaining the project schedule.
- Supports project management and the team as a contributing member.
- Minimum of three years of project scheduling experience."

PMI Agile-Certified Practioner (PMI-ACP^{SM}):
Someone who has experience in the tools and techniques of the agile methods for project management
- It requires a combination of training, experience and an exam. It also bridges agile approaches such as SCRUM, XP, LEAN and Kanban

PMI Professional Business Analyst (PMI-PBA^{SM}):
- Focuses on the experts background with business analysis

- Allows individual to highlight skills in defining requirements from a variety of stakeholders.

The organization makes every attempt to assist individuals in growing their skills to achieve more efficient and profitable management of the projects for which they are responsible. For more information, check out www.pmi.org.

While domestic usually indicates a single home-based country or nationality, we will treat domestic projects and this term as covering both the US and Canada. There are many more similarities working within the US and Canada than differences. Some may challenge that Eastern Canada requires French-speaking skills, but one might also argue that in the US Spanish-speaking skills are beneficial in major cities. Both are romance languages and can be dealt with using an appropriate multi-lingual team. The similarities to which we speak are the general cultural characteristics and demands within a similar region. Acknowledgement is given that this is perhaps a very broad acceptance, but as we delve much more into the internationalization issues over time, we will find this to be a fair assumption within the scope of this book.

One further requirement expected is a basic understanding of the Project Management Body of Knowledge. This is the formal standard provided by PMI for the benefit of unifying project management techniques, terminology and processes. They are formulated into ten bodies of knowledge (BOK) as identified below. We will give a brief explanation of each of these BOK, but would expect that those interested in international project management have some background in managing projects and will seek formal project management training through an authorized organization approved by PMI.

Bodies of Knowledge [15]

✓   Project Integration Management

In short, the goal of project integration management is to unify the various processes within project management in order to assure a successful completion and meeting the quality metrics for the scope, schedule and budget. This process includes all of the tasks from inception of the project charter, which assigns the project manager, to the closing report. In practical situations, this is the function that will also help tie the project completed to a general database so that all future project managers can learn from the successes and challenges experienced by the project team.

✓   Project Scope Management

The ultimate goal of project scope management is this: "includes all the work required, and only the work required, to complete the project successfully." Too many times the definition of the project 23

is rushed because of the typical belief that "everybody knows what is supposed to happen." However, ask one hundred teenagers what should be done to clean their room and you will get one hundred distinct answers from the extremely organized teen to those uninterested in cleaning at all! Making sure that adequate time is dedicated to initially defining expectations that will be included in the work product, whether product or service, is absolutely required. Some projects do fail in later stages, some don't even finish.

✓   Project Time (Schedule) Management

The goals of project time management should be fairly obvious: deliver the project on time. However, to accomplish this goal, individuals must define the tasks to be completed, assume relationships to sequence these tasks, estimate the time to complete each task and formulate a resource schedule that can be met. Too often individuals do not take the time to make the estimates per deliverable or sub-deliverable and a rough order of magnitude estimate is used for the final projection of delivery date. Hence, most projects are late.

✓   Project Cost (Budget) Management
The ultimate goal of project cost management is to stay within the expected budget. However, this should be maintained both on the high and the low end. The temptation always exists to provide a higher estimate to assure enough budget, but what usually happens is that the project will expand to fill and surpass the budget. This is inevitable with the characteristics that most individuals use to manage their projects.

✓   Project Quality Management
For most engineers this is an elusive topic because the separation of product quality from project quality is difficult. The product quality will be seen for the life of the product, while the timeframe upon which the project was delivered will only be remembered for a short period. Project quality management strives to assure that the scope, schedule and budget, among other metrics and bodies of knowledge, are delivered within quality metrics, policies and procedures.

✓   Project Resource Management
The body of knowledge dealing with individuals, teams and the management of these teams is known as project human resource management. The goal is to use best practices to achieve the greatest use of individuals and involvement of expertise to succeed.

14

✓   Project Communications Management

Considering that 90% of a project manager's job is communication-based, this should be an area of great interest. If you ask numerous employees why they feel that glitches happened, it is because they were uninformed or misinformed. Communications covers the tools, timeliness and methods of disseminating information to team members and stakeholders during the project planning, execution, controlling and closing phases.

✓   Project Risk Management

Project risk management deals with the identification, estimated impact and likelihood, and response plans for risk events. Risk cannot be completely avoided in most circumstances, unless the project is avoided. Some risks can be reduced or mitigated, others transferred and some accepted. These decisions cannot be made if the team does not evaluate situations that can go wrong for the project. Construction teams know that weather is a risk and therefore plan for this issue. Every project will have various types of risk and proper planning can help reduce the catastrophic level to a manageable level.

✓   Project Procurement Management

No company, no matter how large or sophisticated, has all of the resources it needs under one umbrella. Every company must use outside resources for securing people, materials or equipment to complete their project. Thus, project procurement management techniques are used to negotiate contracts to secure these resources at the necessary time with company funds.

✓   Project Stakeholder Management

This body of knowledge used to fall under the communication chapter, but since the fifth body of knowledge it has it's own section. The goal is to identify individuals or groups (union, society's, etc) which have an interest in the outcome of a project. The stakeholders are also assessed for power and interest in a project as well as considered for how best to engage them in the project.

✓ Ethics

PMI, in addition to the bodies of knowledge have addressed ethics or professional responsibility. The reason for this is that most accrediting bodies expect the members and particularly those that carry a recognized certification to uphold high values for personal and corporate status. Individuals that carry the PMP must meet the code of conduct or risk losing their certification. One might consider many of the expected practices to be commonplace, but what we will find is that ethical expectations in various cultures, vary dramatically. PMI sets the guidelines for the international body of project managers. The FCPA or Foreign Corrupt Practices Act is one of the methods used to solidify a standard of expectations from foreign locations. Classes have been developed to address this topic and help educate people in a unified fashion.

Project Management		
Project Scope Management	**Project Time Management**	**Project Cost Management**
Scope Planning Scope Definition Create WBS Scope Verification Scope Control	Activity Definition Activity Sequencing Activity Resource Estimating Activity Duration Estimating Schedule Development Schedule Control	Cost Estimating Cost Budgeting Cost Control
Project Quality Management	**Project Human Resource Management**	**Project Communications Management**
Quality Planning Perform Quality Assurance Perform Quality Control	Human Resource Planning Acquire Project Team Develop Project Team Manage Project Team	Communications Planning Information Distribution Performance Reporting Manage Stake holders

Project Risk Management	**Project Procurement Management**
Risk Management Planning Risk Identification Qualitative Risk Analysis Quantitative Risk Analysis Risk Response Planning Risk Monitoring and Control	Plan Purchase and Acquisitions Plan Contracting Request Seller Responses Select Sellers Contract Administration Contract Closure
Project Integration Management	**Project Stakeholders**
Project Charter Preliminary Project Scope Statement Project Management Plan Direct and Manage Project Execution Monitor and Control Project Work Integrated Change Control	Identification Power/Interest Engagement

Figure 1.2. PMBOK Knowledge Areas and Processes

Note that in late 2020 the 6th PMBOK will be obsolete as PMI plans to release the 7th edition of the PMBOK.

One last item to address is the fact that no project can be everything. Just as the opening quote identified, there are three primary criteria and any project can meet two of them. Aim for all three and you will fail! You would do well to keep this in mind.

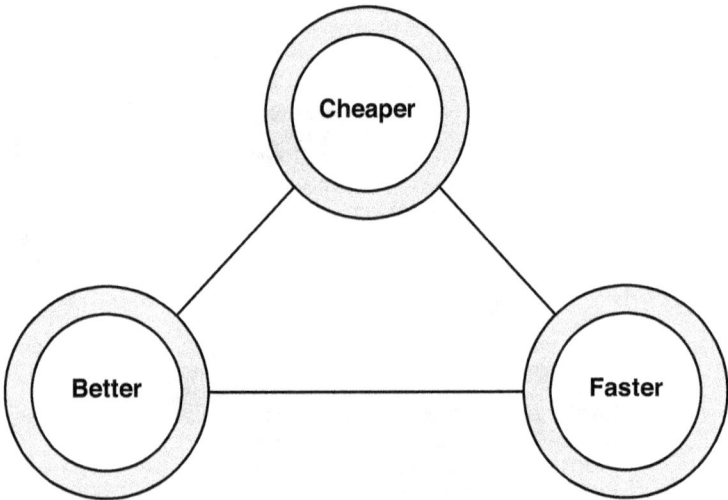

Cheaper

Better

Faster

Figure 1.3. Triple constraint (Scope, Schedule, Cost)

In addition to the triple constraint or the "big three" we also consider quality, risk and resources as constraints too. So, there is much to consider in the way of limitations for a project manager

One other note that we will work through a bit later is that this book is focused on traditional project management which is also called "predictive" or "waterfall" methods. The use of agile methods or hybrid will be mentioned throughout the book. However, they have many additional tools that will not be explored by a predictive environment for construction.

Please note that this text is a survival guide and not a novel. The intent is to share enough technical information and practical experience to help you in your quest to successfully manage international projects. Keep this in mind and look for the practical applications in your projects as examples and information are shared.

Chapter 2
Projects Types

"As we have no experience on this type of big project, huge money will have to be spent on foreign consultancy to conduct the study of the project." (Saifur Rahman) [1]

One maxim that I've used over the years has been that "we must learn from the past, use it in the present to get to the future." For project management this challenge is tough to overcome because past examples are not perfect representations of the present. This is true because projects are unique. However, we must learn lessons from every project in which we participate, cooperate or lead and use these to grow as professionals. The variety of challenges that face each project manager are as diverse and unique as the projects themselves. One would not expect to find the exact challenges in organizing a cross-country band tour as building a skyscraper, but planning, execution and control techniques are all needed regardless of the project type.

Let's first explore what the different characteristics of various projects that might be in store for project managers. Consider the following list:

- Short duration projects vs. long duration projects
- Large projects vs. small projects
- Development of best-in-service, best-in-performance or best-in-cost
- Domestic, dual-nation, global
- Industries to consider:
    - Construction
    - IT

- New product development
- Olympics: Beijing versus London, Rio vs. Tokyo
- Service industry

There are countless categories that might be considered for the opposites in project management, but we will focus on these for the majority of this chapter. The hope is to explore the differences and learning how other types of projects function and challenges just might help reduce stress when similar challenges are faced on your own project within your own industry.

Short Projects versus Long Projects

Depending on personality types, project duration can be a major concern. For those that are the fabled hare, from the *Tortoise and the Hare* [2], they may be thrilled to have a list of short projects versus one longer project. For personalities that favor accelerated performance with periods of free time or vacation, having short projects is terrific. It will lend to learning something new from each new experience and assure that boredom does not set in.

The other perspective of the fable is the turtle that will appear slower yet work consistently. According to the fable, the turtle will win, not because of speed, but because of persistent effort. For projects that will have an extended period, the turtle personality will be more desirable and more effective.

Consider the following example of this in real life:

> *University staff positions require constant coordination, but some are in short sprints and others are in for the long haul. Two particular positions that come to mind from personal*

experience are that of administrative coordinator for online classes and graduate student for a new-to-world engineering project.

The administrative coordinator for a five-week class must coordinate with hundreds of students and make sure that each is aware of the new student orientation, start and completion dates for the class, name of the professor, name of the teaching assistant to whom they will report, how to contact technical support in case of computer issues, where to purchase books and a myriad of other class details for which they alone are responsible. All of these responsibilities are necessary to assure that the students learn successfully, however, all of the details must be coordinated and disseminated within the allotted time. Running over the schedule is simply not permitted. Once the course is completed, there is perhaps a week to sit back and enjoy the accomplishments, but the next class is ready to begin and a new class, new professor, new details of the course, and new challenges a wait.

On the other hand, a graduate student would work with the professor to accomplish a particular development goal. Assuming that this graduate student will be in the engineering program for about four years, they may work with the sponsoring professor to accomplish the goal over that entire period. While the immediate demands of a five- week course may not be apparent, each week progress must be made to achieve both the goal of the project and the ultimate goal of receiving a doctoral degree. If one week is missed, that probably isn't devastating, but it needs to be made up during the rest of the project.

The point of this example is not that one position is better than the other. This is entirely personal preference and individual matter. Some enjoy the project sprints and others prefer not to be rushed, but would rather have a steady workload, albeit, still busy! The relative advantages to having shorter projects are that they tend to have periods of lower demand in between high demand, tend to allow for a variety of experiences and also allow individuals to change project teams and get to know more people. However, the challenges are extremely tight timeframes, many with virtually no slippage permitted, increased risk of budget overruns because controlling costs through arduous estimation processes are not permitted and the variety actually may be stressful since broad expertise is a requirement.

Before accepting one or the other type of project, be careful to understand your own likes and dislikes.

Large Projects versus Small Projects

Another consideration for accepting projects, or at least understanding the project thrust upon you, is that of the magnitude of the project from a resource standpoint. Resources are generally accepted as human resources, materials and equipment necessary to complete the deliverables within the scope of the project. From a magnitude standpoint, we must evaluate each category separately due to their individual challenges.

From a human resources standpoint, if your personality likes to blend in with the crowd and simply perform your function as a part of the team, then a small project team may be uncomfortable. Every individual tends to be extremely exposed and required to step up and be a leader in various areas. Conversely, if you are outgoing and like to be recognized for your work, blending in with a large organization would be unfulfilling and likely lead to some level of dissatisfaction. The larger the project, the less visibility you will have. This includes the project manager who will have less and less contact with the individual team members.

From a materials standpoint, the magnitude of the project can be a challenge simply for logistics and timing purposes. If a construction crew ordered and took delivery on all of the materials needed for a skyscraper on day one, there would be absolutely no room to move! The materials need to be delivered one floor at a time or perhaps staged a few floors at a time. This is particularly true for locations in developed cities such as New York, Boston, Chicago and Los Angeles with no space in the surrounding area to work. Everything from manpower, materials and equipment must be contained within the permitted work site.

Equipment is not much different than the case with the materials, but does require excellent planning. The specific type of equipment, time for the equipment and coordination with skilled labor to operate it all necessitate justification.

Development of Best-in-Service, Best-in- Performance, Best-in-Cost

Regardless of whether you are developing a new product, a new service or an improvement upon existing products and services, you need to understand the ultimate goal of the end work product as one of the three "best" categories. Are you going to be the best in performance, cost or service? This is defined as the triple constraint for *projects* in the last chapter, but it is a little more defined from a *product* or marketing standpoint here. Usually a company or project team will decide what the ultimate goal is within the mission or scope.

25

A book written by Treacy & Wiersema called the Discipline of Market Leaders [3] recognizes that an organization must commit to one of these three principles to set themselves apart from the competition. The three can be prioritized from one to three for any given company, and no project or company can ignore the other categories, but one must be prevalent. Examples that were given in the book include:

> *Best Performance*: Sony as a product leader that always has the first product to market.

> *Best Service*: Airborne Express (purchased by DHL in 2005) which always delivered for their customers.

> *Best Value*: McDonald's is the obvious selection; they focus on low price, speed and cleanliness, not a premium burger.

Once the company has a focus, this should be carried down to the project level. If you cannot tell your focus for the specific customer, then ask.

Seek out an answer and if the executives of your company cannot decipher either, have them read the Discipline of Market Leaders. It may be more than a couple decades old, but it pinpoints why many companies fail when there is a squeeze in the economy.

Domestic, Dual-nation or Global Projects

The focus of the remaining chapters of this book will be on the specific aspects of international project management and the challenges, stories and methods to overcome the obstacles. Understanding how your current project or future projects deal with internationalization issues will help to reduce stress and project friction points.

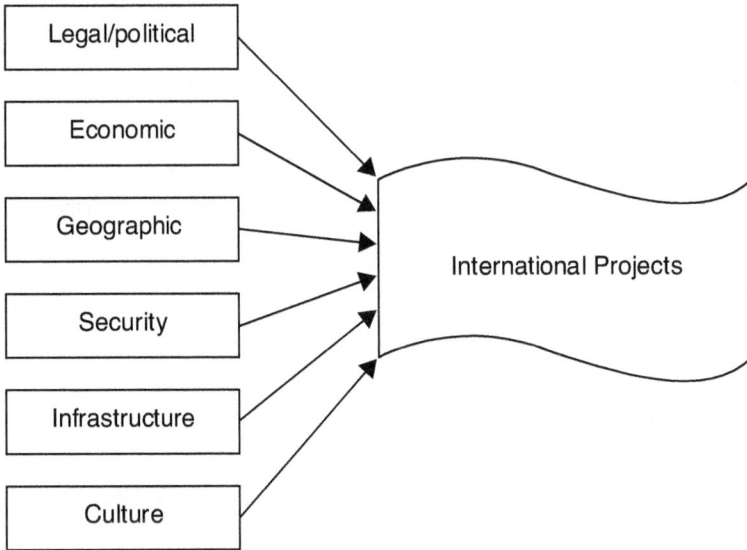

Figure 2.1 International Factors Affecting International Projects [4]

Industry Differences

Depending on the specific industry in which projects are based, there are different demands and perhaps greater or lesser planning effort required. The organization may need extensive control mechanisms to assure failure is not a possibility or perhaps the costs of control are more than the costs of rework. The specific details of the project are of course in the hands of the project manager and project team. Thus, experience and decision-making skills are as important as understanding a uniform process. Together, the skills and processes will lead to project success in any industry.

Construction

One of the initial industries in which planning was crucial had to be the construction industry. Consider back to the great pyramids in Egypt and the architects of that period. The precision with which the stones were laid was no accident. Somebody envisioned a great project (initiation) and took the time to plan out the size and scope of the pyramid with great effort (planning). They used significant amounts of labor (execution) to build the main portions of the structure, continued to monitor that all stones were placed properly or corrected, if not (control) and ultimately completed the project and dedicated the pyramids.

In more modern times the same processes were and are used to construct buildings from Notre Dame Cathedral to the Taj Mahal and ultimately skyscrapers such as the Sears Tower, Taipei 101 and the Burj Khalifa.

With in this industry, great effort had to be instilled to assure that physical safety for the construction crews and ultimately the

occupants would be of utmost importance, but each of these structures also incorporated aesthetics and architectural nuances too. As a note, each of these structures is also in a different country which have different local regulatory bodies. Also, weather conditions stipulate unique building principles and requirements. These may fall into technical expertise, but project managers must have some familiarity with the technical requirements (within the project scope) to deliver a successful project.

New Product Development

The area of development has been around since the first wheel was invented. Even prior to the wheel were basic tools such as bowls, hunting spears and the like. In primitive times imagine the craftsman that was capable of developing a spear that would fly farther, straighter or with a more durable point. They would be invaluable to the family circle or community in which they lived. They were the first product innovators.

Over time, individuals have designed and built inventions from carts to cars, paper airplanes into hypersonic jets, hand tools into power tools and building blocks into complex toys such as Kynex and Lego sets. Today most homes in North America rely on a switch to power lighting, a knob to get fresh water and a key, steering wheel and two pedals to move them at high speeds from one place to another.

Beyond these "essentials" we also have radio and television to act as informative communication tools, but also allows us to explore the world through cultural programming and so-called entertainment!

New product development requires a level of creativity to take a spark of inspiration and turn it from an idea into a functional device. Once that idea occurred, the same processes of planning, execution, controlling and closing must be completed to implement the new idea. Unlike construction, which generally requires a variety of expertise to complete, new product development might be completed by a single individual for a simple to moderate device or hundreds of participants, depending on the level of complexity.

Products have a life cycle and so do projects. Refer to Figure 2.3 for a comparison of the two different cycles, which have some similarity and of course some differences as well. The project manager who is a product manager as well has to be able to wear two hats at once and know which hat is suitable for the occasion.

Figure 2.2 Pick your hats wisely

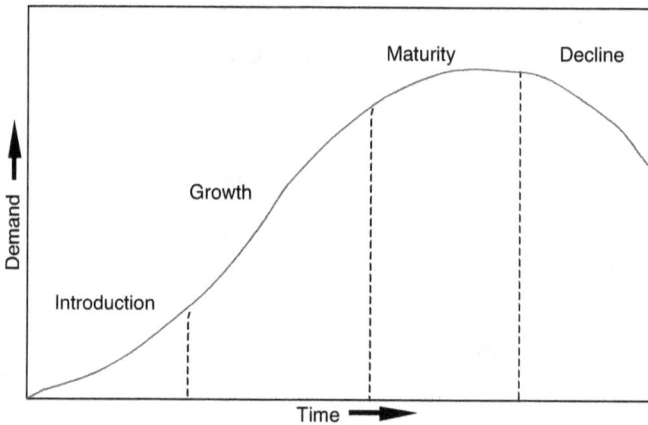

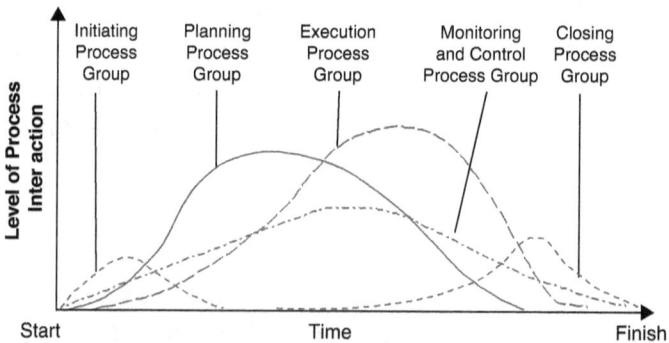

Figure 2.3 Product Life Cycle (top) [5] and Project Life Cycle Process Groups (bottom) [6]

Information Technology (IT)

The computer age brought on new requirements for individuals that could program the digital machines for different functions. At first glance it might appear that the complexity of projects encompassing programming computers is more difficult than construction or new product development, but this is due to the technical aspect. Depending on your perspective, perhaps it is easier or more difficult, but this is irrelevant. The processes will be the same, but the application of the processes will be different.

The idea back in 1963 for the ARPA team led by Joseph Licklider [7] was to incorporate communication between computers and ultimately allowed separate locations to transfer data. Beyond this initial communication the project expanded and further sites were included. The technical expertise was matched by the project coordination.

A favorite movie quote was in the movie Jurassic Park [8] when the character Dennis had programmed the cameras and security system to intentionally malfunction so that he could steal dinosaur embryos. The comment from Mr. Hammond, the project manager, was that "when Disney world opened there were all sorts of problems." Dr. Malcolm's response was "yes, but when the Pirates of the Caribbean broke down, the pirates didn't eat the tourists." Why is this relevant? Obviously because computers create or prevent safe conditions for the users. Similar to construction and product development, information technology has an extremely important safety aspect in the planning, execution and controlling aspects. Perhaps this was an extreme example, but security of credit card and personal information is paramount in IT. Clearly with the data breaches at major companies over time, it is still a challenge.

Olympics – Beijing 2008 vs. London 2012

In 2008, Summer Olympics Games in Beijing were a spectacle unlike any prior. The opening and closing ceremonies were likely the most complicated performances in history including tens of thousands of performers, each clad in specific outfit and expected to hit their mark during the performance with extreme precision.

Beyond the opening ceremony there were the logistical efforts, catering and of course the preparations that were made for the venues themselves. The Watercube was a modern marvel that uses 90% solar power to heat the pools and the Bird's Nest allowed for natural ventilation. [9]

Considering the magnitude of a $42 billion effort [10] that would take place in a mere 16 days, the planning had to be superior. There was no time to make up in case of a mistake. All of the issues had to be addressed in advance or the events would be delayed or worse yet be cancelled. You cannot have a swim event without the pool being filled with water, right?

When preparing for an event, as opposed to other types of projects, the planning is essential because the project is time- constrained, the execution has no slack and controlling must be done in virtual synchronization with the event itself. So what was the difference with the Beijing 2008 Summer Olympics Games and the London 2012 Summer Olympic Games? London noted at the 2008 Beijing closing ceremony that they would be on a much tighter budget. London's organizers made it clear that they could not match the magnitude of the event in China. Not only would the London games be time- constrained, but they would also be resource-constrained, which means that the scope has to be reduced.

We can also consider the challenges faced by the 2016 Rio games and the now-delayed 2020 to 2021 Tokyo summer games. Rio experienced many challenges right up until the events were starting. Many of the dorms for athletes were not ready on time and the plumbing actually was not working in some places. Those games have been considered some of the worst planned in modern times due to lack of planning and even more criticism of the government because the venues were not used after the fact. One year later, many photos were released of the venues covered in vegetation and basically in ruins. In contrast, Tokyo was up to the challenge for 2020, but then the COVID-19 scenario hit and for the first time, the modern games were delayed to 2021. Other games had been cancelled due to war (1940 and 1944), but these games have been pushed back for a year and require complete revision of schedule and budget. Scope cannot be ignored either as additional health monitoring will certainly be considered, even a year after COVID.

Service Industry
If we didn't address the specific needs of the service industries, we would be negligent. Projects to deliver a new service to the market now encompass the majority of revenues in the US rather than goods. [11] Interestingly enough, the costs of the projects to deliver the services is usually less in dollars and time than for physical products and they have higher margins. The challenge is that in times of economic hardships, people learn to live without the comforts of certain services and this area shrinks at an equal or greater rate than corresponding goods.

Delivery of a service project might best be given in that of the cell phone market. While the cell phones are becoming more complex and detailed in performance, they are becoming give- aways, provided to the customer if a commitment is made to the service offer. The latest and greatest phones typically are offered free or at a significant discount with a 2-year service contract. This transforms the service into the actual offering and the product is a perk. Cell phone service offerings are usually one of three things:

- Best cost value for the minutes received
- Best customer support in-store and on the phone
- Best network connections (quality and location availability)

These value propositions take us back to the first chapter. The same market leadership characteristics are available from a service such as a cell phone and the associated products. Defining the scope of the service, planning the offer, execution of the offer through promotion, advertising and direct sell efforts *are* the project. The expected outcome is increased revenue through service contract guarantees.

It should be evident by this point that there are many different ways to segregate projects and the differences are basically in the technical elements of the project. Whether you are working with a small or large project, one of short or long duration, local or global expanse, or regardless of industry, the same basic bodies of knowledge and initiation, planning, execution, controlling and closing processes must be completed.

However, the focus of the book is upon the international differences and the special needs of projects that require dealing with multiple cultures, multiple languages, time zones and a myriad of challenges that you might encounter in these projects.

If what you have read thus far is confusing and some of the terminology is foreign, you might consider reading a basic text on project management or consult the Project Management Body of Knowledge published by the Project Management Institute. If you are comfortable, then let's proceed to the good stuff!

Chapter 3
Cultural Bearings - Social

"Americans who travel abroad for the first time are often shocked to discover that, despite all the progress that has been made in the last 30 years, many foreign people still speak in foreign languages."
(Dave Barry) [1]

For those that heave never traveled internationally, lived in a multi-cultural family or have somehow averted meeting individuals from different cultures, this could be somewhat of a shock, but landmarks are not the only differences in foreign countries. By estimation, those that have lived in the US have lived, worked or discussed issues with those from another country. Whether the conversation went well or not depends on how much willingness both parties have to developing a deeper understanding of the cultural roots in each of the nations.

Many of the issues that occur at the initiation phase of an international project are based on a lack of understanding for the cultures involved. Many false pretenses such as "all people from India are smart" and "all people from Mexico are lazy" create unrealistic expectations or apprehension. This simply is not fair and creates fear. An acronym used to explain fear is:

FEAR = False Expectations Appearing Real

Too many times it is the unknown that will drive us to believe unreal stories, drive perceptions beyond investigation and ultimately creates the stereotypes by which we live. Giancarlo Duranti

35

recommends "Learn from generalizations about other cultures, but don't use generalizations to stereotype." [2] The best way to learn about a culture in which international projects will function is to experience it for yourself, but short of a 3-month stint in each of the countries involved, a thorough study of the country may have to suffice.

Let's address the challenges that will be involved with international projects one at a time from a social standpoint and work from there. This may seem a bit like a travel-channel special, but we will try to introduce some of the business aspects into many social circumstances. Remember that before people become business oriented or project oriented, they are first people!

Language Barriers

The assumption for this book is that the reader is from the US or potentially Canada and thus English is a primary language. Perhaps Spanish or French are also fluently spoken, but these are the three North American primary languages for the last two hundred years. For much of the US and mid- and western Canada though, English is the only language spoken in the household or at work and thus a new language is intimidating and well, quite foreign. It might take some time to adapt, but to work with international projects, it requires a level of effort of both parties, even if English will be the selected language for correspondence.

Already twenty years ago I participated in a training session where all of the training staff for a 40-hour seminar were from the Midwest and all of the participants in the class were from Texas, Alabama and Louisiana. The first day was difficult for both parties simply due to the fact that each carried an accent from their area and the individuals had to decipher what "y'all" meant or "yeah" meaning yes and not "go horse". It was quite entertaining to see and be involved. By the end of the week one of those from the Midwest was speaking with a southern drawl. This is not brought up to be derogatory to either the Midwest, South or any other territory but to bring awareness that even territorial accents within the US can create a challenge, if we don't speak clearly and with diction.

Now, add onto this challenge the fact that North Americans are extremely fond of slang and catch phrases, which do not mean the same thing in foreign countries. Some of the misconceptions include:

Here are a few examples I've run into:

• Knock off	• My car was "totaled" in the accident
• Deadlock	• Things should cool down in a day or two
• Put it to bed	• Don't get so bent out of shape
• Spot on accurate	• Let's go to a bar and down a few beers
• My little ones	• Free and clear
• Touch base	• Let's take this offline (on a virtual call!)
• 50-50	• Back to the drawing board
• Game plan	• Think outside the box

If you are from the US, these are likely a part of your everyday language or at the very least, you can figure them out, but it is not formal or accurate language. Instead, individuals are confused and were quite concerned by the "put it to bed". During visits to the US we had some fun teaching other less business-oriented language such as "you rock" or "that's dope". Don't worry, the humor went both directions on the team!

Even for the poor souls who understand English, they will have a challenge to keep up with the native speakers who use these phrases. Be sensitive to those that are not native English speakers and realize that you have an advantage and maintain courtesy. If they are making the effort, they are sincerely doing you a favor.

Next, realize that translation is a challenge that will require you to focus. The quote that opened this chapter made me laugh and hopefully it is clear that this perception still exists. For much of western Europe, English is a required second language, but that is not the requirement everywhere. Should you be the only individual not to speak the native language, there is a good chance that you will have to be translated for those you are working with. Not a surprise, but working with translators takes time. For various purposes, I've been translated into Spanish, French, Chinese (Mandarin and Cantonese), Swahili, Romanian, Hungarian, German, Russian, Thai, Japanese and Korean. There are probably others, but these are the languages that first come to mind.

One story that seems relevant at this point is that there were Japanese professionals that were sitting in a room and used a young lady to translate for the entire morning of meetings. When it came time for lunch, all of a sudden in perfect English the lead Japanese businessman asked where we wanted to eat. It appeared that they were playing a joke with how clearly they spoke English. Yet, upon returning to the office for afternoon discussions, the young lady again translated for the entire afternoon. Curiosity led to asking why they did this with such clear and capable English skills. The answer was two-fold. First, in case there were some words that were not clear, they would be absolutely certain that the message got across. More importantly though, they found that Americans expect quick answers and hearing it in English, the translator bought time to think about the answer. Also, through the translator, things could be smoothed out. Be prepared for language and culture issues such as this.

SPEAKING A DIFFERENT LANGUAGE

Here are simple rules for formal communication that can help young workers avoid embarrassing and potentially career-damaging blunders.

> **Watch and learn.** Study how veteran employees communicate-and then follow their lead.

> **Target your style.** Leave the text talk, casual abbreviations and poor grammar for chats with friends.

> **Tie your ideas to the company's needs.** A great solution is one that's aligned with the strategic goals of the business and the client.

> **Be proactive.** Don't point out a problem unless you have a solution that you're ready and able to implement.

> **Back it up.** Have facts and figures to support how your idea will directly impact project success.

> **Seek out constructive criticism.** Find a mentor and get feedback on your communication style and performance.

> **Just ask.** It's better to pose what seems like a stupid question than to risk not finding out the answer.

Figure 3.1. Speaking a different language [3]

Note

DuoLingo, HelloTalk and Babbel are three of the most popular apps for learning the basics of a language. In a few minutes a day, you can learn some of the most required phrases and these are all for free. iTranslate, Google Translate and others have live features to help automatically keep a conversation going. Just make sure your device has the processor speed to keep up!

Time Perception

Americans tend to work on a clock and strictly observe on-time and lateness based by the minute. Even within American businesses there is some flexibility to the rigidity of how important time constraints are. A meeting scheduled within a department for a specific time may not raise any issues if one or two individuals are two minutes late. However, a presentation or a meeting with multiple tiers of employees raises a different expectation level that at least those entry level positions are early, managers are on time and executives are right on time.

If you travel for business, be prepared for different levels of preparedness based on the clock. Some cultures remain very rigid on time schedules, but this still tends to be a more western tradition. Once, a visit to Eastern Europe yielded the opportunity to visit a little country village. Roughly 100 of the 120 people from the village attended the single church there. The doors opened at 10am for the devotions and the regular service began at 11am with a traditional 12 pm completion. However, the village farmers were frequently there before 9am for some and others didn't show up until 11:30 am. This seemed odd considering my traditional expectation was a fixed start and end time. Everybody was expected there on time. The minister laughed and related to me that almost none of the farmers had watches and only half had clocks. Many knew the time throughout the day only because the TV stations told them or the location of the sun. This was in the 1990s, not the 1890s!

Be aware that time sensitivity for meetings may be very precise in some cultures and may be more flexible in others. Sober advice is to be on time yourself, but be flexible and open to those that are late. It will give you time to soak in the culture and prepare at a relaxed pace. Buying watches as gifts might be advisable, but don't expect a change in on-time performance unless the individuals work directly for you! This is true for on-site meetings, phone or video conferencing alike.

Currency / Inflation / Income

Of course, if you are going to be traveling to a foreign nation, be aware of the currency exchange issues and understand that while it may look like "toy money" it is quite real. The challenge of remembering which bill to pass is usually the least of concerns. After a short while the numerals are sufficient and adapting to the conversion rate that "this bill equals an American twenty" should become a mental note.

However, under significant periods of inflation, you might find that you need to monitor on a daily basis what value your currency is worth. During a trip to Eastern Europe in the 90s, I learned that post-communism was good to the American dollar and that my first $100 lasted me two weeks! However, after two weeks in the country, the currency had devalued 18%. It was almost unfathomable from my past experience. It reminds me of a quote from the movie *Eurotrip* [4] that a hotel staff member got a tip of a nickel, turned to the hotel owner, slapped him and said "Look! A nickel! I'll start my own hotel!" No disrespect is intended, and maybe it isn't quite that good of an exchange rate, but attention to the detail is obvious.

Also consider the GDP and per capita income of families. This will give a general picture of how wealthy project team members may or may not be. Consider the relevance of topical discussions when one family is fighting to feed their children, while others determine which is the better sport car to buy. Americans tend to share these "concerns" but it may only drive a wedge between the foreign counterparts who may or may not be compensated as well.

For an international project the greater benefit to take from this point is that if you plan budgets, plan for inflation or deflation to affect your budget dramatically. If you are building products, prepare for product costs to increase or decrease, travel budgets to be affected and salaries for employees to fluctuate. A consistent upward or downward trend is easier to accept than the constant fluctuations. This fact alone has deterred some companies from working with specific countries. The volatility is enough to make them avoid the headaches.

Food and Eating Patterns

For those that will travel internationally, one particular area of intrigue as well as fear might be at mealtime. Understanding local customs and the foods that might be encountered should help reduce the stress of the unknown. Take some time prior to traveling to learn your way around the table and the likely delicacies.

Be willing to try certain foods that you've never eaten before. This can be a challenge with food allergies, religious restrictions and personal limits of gag reflex. In some areas of China the delicacy is roast scorpion and in others it is water cockroach. Not my idea of fine dining, but others find it a great new treat. However, the effort to try new things will frequently demonstrate your willingness to learn a different culture and helps build rapport with your international associates. At one dinner the rotating table selections included three to five types of fish, snake, and other rarities in meats, fruits and carbohydrates. I've also had my local guides "trick" me into eating some items such as bull testicles! They had a laugh about that one!

One of the other dining challenges is the topic of when meals are taken. In the US, typically there is a breakfast ritual, which for many is toast and coffee. For others a full meal of eggs, bacon and juice is the expectation. There is a typical lunch hour somewhere around noon and dinner is generally between 5pm and 7pm. Not everywhere, but this is a good rule of thumb. Other cultures eat more, smaller meals in a day, some as late as an official 10pm dinner.

During one trip, that lasted three weeks, it was tradition that the business hosts serve some form of pastry and coffee. Also, it is tradition that you must take from the plate more than once or it offends the host or hostess at the meeting. This particular trip required meetings every two hours at a different location from 8am to 8pm or later. This meant three or four meals and typically five pastry sessions each day. The food was wonderful, but added more than fifteen pounds to my waist in those three weeks. On my next trip there, it was wise to take a very small portion and schedule fewer meetings or make sure that my translator indicated that we had just eaten lunch and that the businessman is very busy so he may only take once from the pastries during the meeting.

Make sure that you set expectations clearly if you have allergies or health related demands and "eat respectfully" around those guidelines to make sure you remain a pleasant guest. Bon Appetite!

Security

Security is another area of personal interest. Depending on the country or countries to be visited for the related project, there are varying degrees of protection that must be taken. Even when traveling to other countries with fairly wealthy populations, there are risks inherent to travel. Individuals that wear expensive suits, watches or even shoes mark themselves for pick-pockets. Losing your wallet can be a tremendous setback while in another nation not only because you are without money, but possibly credit cards and identification. This could be a great detriment to your trip and will at least delay your plans.

On the other end of the spectrum is the potential for kidnapping. In certain areas of Mexico and South America it is unwise for anyone not from the area to walk around by themselves, even during the day. It has been related that individuals are thrown in the trunk of a car and forced to use their credit cards at each ATM the bullies force them in front of, and then eventually beat and dumped by the side of the road. This might be an extreme and rare case, but it would indicate a certain level of wisdom when in these areas and not to walk by yourself. Work with the local contact, whether employee, employer or business associate to help determine what steps are wise. However, a few recommendations are to keep your cash in more than one area (belt, shoe, wallet and other locations), keep more than one credit card and in more than one place on your person and of course always be watchful for suspicious behaviors.

Check out statistics for safety in Columbia and you will understand the point here. [5]

Transportation

Another area in which some caution must be taken: transportation, both people and materials. Considering that the US has the best transportation infrastructure of any major industrialized country, this would indicate that nearly any other part of the world could be a potential disappointment. Of course, the roads and bridges in the United States require constant upkeep and after the August 1, 2007 collapse of I35W in Minnesota, awareness to such improvements increased nationally.

For a leisure traveler or project manager and employees that will be involved directly in travel abroad, make sure to get information on the road conditions, driving and general laws and guidelines for the area. One recommendation already made thirty years ago was that if you plan to rent a vehicle in the main cities of India, make sure to take all of the additional insurance you can get. [6] Many times the cars in the big cities will drive six inches apart at 60 miles per hour and unless you have the nerves of steel to handle this, it would be advisable to hire a local driver for your stay. Another experience similar to this was a stay in a third world nation where we luckily had hired a driver. On the second day in the city we were hit from behind. The drivers got out and spoke calmly for a few minutes and while we were sitting in the car, another vehicle side-swiped our car and took off the side mirror. In astonishment I said "he's getting away!" My translator who was from the area calmly replied, "not all drivers stop." With this understanding, some words of warning: make your decisions on driving with care.

Another aspect of transportation is with the materials and finished goods for products that are produced in other nations. One issue as already noted is the poor conditions of many roads. This would indicate that packaging must be made with great care. Frequently trucks skid off the road and bump rails or nudge one another and the materials and goods within must be able to handle such jostling without sustaining significant damage. This may seem extremely suggestive that all foreign drivers are poor, which is not the intent, but rather that caution is used wherever you go. Too many poor conditions add up to risks for a project that preparations for such risks must be made.

One experience related to this was the 12" snowfall in the early months of 2008 in China. The areas surrounding Shanghai came to a standstill. American companies were told to expect at least a week delay. This seemed excessive considering that Chicago, New York, Minneapolis and Boston are all hit by such snowfalls each winter and continue to operate, maybe with a few hours' or perhaps day's delay. However, China does not have heavy equipment to remove snow from the roads. The projects, materials and individuals were essentially stuck for a week. Plan for the worst and be aware of the risks.

One last thought on transportation that was in the news in late 2008 included the pirates off the coasts of Somalia. Some CNN news reports at the time identified that local politicians noted the pirate "business" as the greatest source of revenue in the nation. Both products and individuals are at risk of being taken hostage for a ransom which in 2008 was more than $30 million dollars [7] This again might be an extreme case, but kidnappings and hijackings are a real threat world-wide.

Religion

Note that the differences in religion will affect work in various projects. Individuals that travel from the US, which is tolerant of thousands of various religions and sects, is the anomaly in world faiths. China prohibits public worship, being a communist state, and various other countries still declare a national religion, even if others are permitted. Regardless of your affiliation or lack of affiliation with a religious body, the rules and guidelines of the local traditions and regulations must be adhered in order to sustain cooperation. In the Middle East, the Western oil companies must respect the prayer rights of the employees at specific times of the day. While teaching courses in the middle east, I adapt the breaks to the prayer times to respect the local customs.

In certain nations individuals are expected to wear the local garb in order to respect the traditions and therefore women must not be seen wearing risqué clothing or uncovering at all. Such disrespect is seen as a crime and women will be jailed or executed for failing to meet the guidelines. [8] Education of the specific country's guidelines is, of course, a necessity.

Holidays

We all enjoy some time off, whether you are willing to admit it or not. For the United States it is expected that a few extra days are taken off around late November and December for the national Thanksgiving holiday and Christmas, Hanukkah, Kwanza, winter solstice or new year. We expect the extra holidays and enjoy this time with our family. Usually we get one or two days specifically for the holidays and additional vacation time is taken, if any is left. This doesn't stop business, but rather slows manufacturing and corporate environments while accelerating retail and some service businesses dramatically.

Over the years it has become apparent to me that the US cannot stop some Swedish companies from taking four weeks off in the summer and four weeks off in the winter for their vacations. Some entire companies are shut down. Plus, they each get four weeks additional vacation. Certainly not all Swedish companies work this way, but some of my projects have had to adapt to this environment. Quite frankly, I'm jealous of the amount of free time they receive!

However, one particular example of holiday challenges for project managers is manufacturing in China. Our projects were slated for late February completion and after the first six months of delays, we figured we would catch up in January after the US holidays and be back on track by February. What a surprise when the Chinese New Year required us to accommodate an extra three weeks' time into the schedule! We were notified of the eight-day holiday two weeks in advance. However, we were notified that the plants shut down at least three days in advance so all of the employees could travel back to visit their rural families. Eight days for the holidays and three more days for returning to the cities must be accounted for. Then we had to account for ramp down and ramp up time. We were an additional three weeks behind schedule. For the next project, we were well prepared to avoid shipping container shortages in September to November from China and the holiday cram from December to February. With the known work outages, plans could remain on schedule. (See Figure 3.2) Plan for the local holiday schedule at the very beginning of the project!

Family Life

Social customs in personal life affect the cultural behaviors in both social and professional situations. Adjusting your own personal views to the common practice of the culture with which you will interact will allow for a more pleasant and open discourse. In China, it is typical to have only one child due to the financial penalty of having more than one child. As of 2006 the cost to have a second child was somewhere around $30,000 US [9] and for a third child $300,000 according to an associate from China. However, by 2016, China began to ASK couples to have two children to help offset the aging population. However, by 2019, China's birthrate was the lowest since 1961.(https://www.bloomberg.com/quicktake/china-s-two-child-policy)

If you work with those in East Africa, the expectation is different. Many households had eight to 13 children, despite the economic burden. Western families tend to have two or three children. There are various social, political and economic reasons for the variations in number of children in the household, but be aware of the fact that individuals are typically proud of their offspring whether they have one or thirteen or more.

Another aspect of the family relationships may be the reliance of one generation upon another for sustenance. In less affluent societies the first generation, we shall call them grandparents, frequently have all of their debts paid and work for cash to save for retirement and help their children, we shall call them parents, to achieve success in their career and work to pay off their debts. The middle generation, "the parents" has children as well and they rely upon the grandparents to feed, clothe and school them. The grandparents frequently rely upon the parents for medical attention, running errands such as food and making sure repairs and such on their homes or apartments are attended to properly. Therefore, it makes the professional-age generation responsible for both their parents and children on top of the career with which they operate. This will put great stresses on them to perform in all areas of their life. Sensitivity to this element of your foreign project team will help you to understand the needs for flexibility and understanding.

Specifically I can relate this to the social structure in Romania in the 1990s, post-communism era. The grandparents (first generation) who traditionally lived on farms would provide food to the parents (second generation) in the city. The second generation in the city would have an apartment for which the first generation had paid and then work their entire life to provide money to their parents (first generation) and save for an apartment for their own children (third generation). This third generation would learn as they grow up that some day their parents would need their help and they would be responsible for both the older and younger generation.

One major shift that occurs as a society becomes more wealthy is that the younger generations have less and less reliance on the older generations. Instead of the need to have parents buy the apartment or home, they can get a loan and pay for it themselves. This might seem freeing to the third generation, but it creates social challenges for the parents and grandparents who have traditionally relied upon the next generation. Beyond the fact that the third generation also loses reliance upon the older generations, they also tend to be better-educated and more open to new business ideas and thus the tides have turned where the youngest generation owns the private businesses and has secured higher level corporate positions. The older generations work for them. Family ties are breaking or maybe better said, fading, because of this. Nothing can be done to reverse the bonds of family ties by you as a project manager, but being aware of these challenges might help you to understand the needs of employees better.

With all of these social concepts in mind it should make you more aware of the need as a project manager to gain a clear understanding of the project communication and project human resource elements. You can do very little to influence the personal preferences and impacts on business from these aspects, but as the saying goes "awareness is half the battle."

2021

Holidays & Observances

January

Su	M	Tu	W	Th	F	Sa
					1	2
3	4	5	6	7	8	9
10	11	12	13	14	15	16
17	18	19	20	21	22	23
24	25	26	27	28	29	30
31						

February

Su	M	Tu	W	Th	F	Sa
	1	2	3	4	5	6
7	8	9	10	11	12	13
14	15	16	17	18	19	20
21	22	23	24	25	26	27
28						

March

Su	M	Tu	W	Th	F	Sa
	1	2	3	4	5	6
7	8	9	10	11	12	13
14	15	16	17	18	19	20
21	22	23	24	25	26	27
28	29	30	31			

April

Su	M	Tu	W	Th	F	Sa
				1	2	3
4	5	6	7	8	9	10
11	12	13	14	15	16	17
18	19	20	21	22	23	24
25	26	27	28	29	30	

May

Su	M	Tu	W	Th	F	Sa
						1
2	3	4	5	6	7	8
9	10	11	12	13	14	15
16	17	18	19	20	21	22
23	24	25	26	27	28	29
30	31					

June

Su	M	Tu	W	Th	F	Sa
		1	2	3	4	5
6	7	8	9	10	11	12
13	14	15	16	17	18	19
20	21	22	23	24	25	26
27	28	29	30			

July

Su	M	Tu	W	Th	F	Sa
				1	2	3
4	5	6	7	8	9	10
11	12	13	14	15	16	17
18	19	20	21	22	23	24
25	26	27	28	29	30	31

August

Su	M	Tu	W	Th	F	Sa
1	2	3	4	5	6	7
8	9	10	11	12	13	14
15	16	17	18	19	20	21
22	23	24	25	26	27	28
29	30	31				

September

Su	M	Tu	W	Th	F	Sa
			1	2	3	4
5	6	7	8	9	10	11
12	13	14	15	16	17	18
19	20	21	22	23	24	25
26	27	28	29	30		

October

Su	M	Tu	W	Th	F	Sa
					1	2
3	4	5	6	7	8	9
10	11	12	13	14	15	16
17	18	19	20	21	22	23
24	25	26	27	28	29	30
31						

November

Su	M	Tu	W	Th	F	Sa
	1	2	3	4	5	6
7	8	9	10	11	12	13
14	15	16	17	18	19	20
21	22	23	24	25	26	27
28	29	30				

December

Su	M	Tu	W	Th	F	Sa
			1	2	3	4
5	6	7	8	9	10	11
12	13	14	15	16	17	18
19	20	21	22	23	24	25
26	27	28	29	30	31	

Date	Holiday / Observance
Jan 01	New Year's Day
Jan 18	Martin Luther King Day
Feb 12	Chinese New Year
Feb 12	Lincoln's Birthday
Feb 14	Valentine's Day
Feb 15	President's Day
Feb 17	Ash Wednesday
Mar 14	Daylight Saving (begin)
Mar 17	St. Patrick's Day
Mar 20	Vernal equinox (GMT)
Mar 28	Passover
Apr 01	April Fool's Day
Apr 04	Easter
Apr 13	Ramadan begins
Apr 21	Admin Assistants Day
May 09	Mother's Day
May 23	Pentecost
May 31	Memorial Day
Jun 14	Flag Day
Jun 20	Father's Day
Jun 21	June Solstice (GMT)
Jul 04	Independence Day
Sep 06	Labor Day
Sep 07	Rosh Hashanah
Sep 22	Autumnal equinox (GMT)
Oct 11	Columbus Day
Oct 31	Halloween
Nov 07	Daylight Saving (end)
Nov 11	Veterans Day
Nov 25	Thanksgiving
Nov 28	Hanukkah begins
Dec 21	December Solstice (GMT)
Dec 25	Christmas Day
Dec 26	Kwanzaa begins
Dec 31	New Year's Eve

Figure 3.2 US Calendar with Holidays

Chapter 4
Cultural Bearings - Business

"The only English words I saw in Japan were Sony and Mitsubishi."
(Bill Gullickson) [1]

We will now transition to the cultural demands that relate more specifically business world activities and not exclusively the social aspects that a tourist might encounter. Individuals that will work with others internationally can take into account the variety of challenges already identified and appreciate the challenges. Each and every difference between cultures offers new obstacles to avoid and new lessons to be learned. It can be exciting or nerve- racking depending on the preparation that is made for the discussions that will ensue.

We shall continue looking at each of the specific angles that might be a challenge and break them down briefly so that you at least have something to think about related to each topic.

Dress Code

The first area of interest is the dress code for your environment, assumed, within the US boundaries. Consider the differences from 1995 to 2005 within US professional offices. In 1995 everybody within the engineering office was wearing a tie and understood that this was not only a choice statement to be dignified in one's profession, but rather, it was a requirement to maintain an acceptable performance review. In other words, it wasn't a choice but a requirement at that time. Something dramatic changed over that decade that permitted individuals to have casual Friday. Wearing jeans was then considered

OK and the most daring of individuals started to do it. Eventually most of the organization followed suit and Friday's were casual. Shortly thereafter, ties were no longer required throughout the work week. Dress casual became the phrase of the day. Next in the line of progression was the coup of dress casual. Companies began permitting the professionals to wear casual clothing every day as long as it didn't distract from getting work completed. Where I was working at the time it took less than two years for this entire transition to occur. The chemists cheered the fact that they didn't have to destroy ties every week by standing over a vat of chemicals and the tie slipping out of the neatly tucked shirt into a hot fluid tank! It was comical to see at least one tie destroyed per week by someone in that department. In fact, for a while, there was a collection of ruined ties for the group!

Where you work may be different. Perhaps there is a dress code that still strictly observes suit and tie yet to this day, but this seems few and far between today for the companies which I have visited in the US. Other companies may require a uniform or perhaps simply permit jeans and a collared shirt. The point here is that even in the US, dress has changed dramatically in recent history and these changes caused friction or nerves for management and employees alike.

Working internationally will offer new insights. In July of 2008 it was my pleasure to teach a course in project management in London to a small group of engineers. The first day is always an interesting event since individuals do not know what to expect from the instructor and it is always interesting to see who attends the courses. In this particular class of nine individuals, eight were wearing ties and jackets. Only one had come without a tie and it was obvious to me that he was extremely uncomfortable during introductions. I mentioned to the class that everyone was free to remove their jackets and to my surprise, none removed the jacket.

The individual without was so uncomfortable that they actually apologized to the others in the group for not wearing suitable attire! What a difference from the US courses. Having taught the same course in the US over four years, there was not one occasion on which I recall a student wearing a tie or jacket into the first day with the exception of a few young ladies who wore a pants-suit.

On the flip side of the coin was a visit to manufacturing engineers in China. I was representing a billion-dollar company to inspect the shoddy manufacturing at a number of equipment and component manufacturers. They were aware of our impending visit, but not one was so inclined to wear anything more than blue jeans and a collared shirt. They were comfortable in their ways and were not threatened by attire. In advance, this topic had been discussed with my translator and they indicated that it would make sense that I pack in the same manner. My outfit usually consisted of black jeans and a dark collared shirt which leant well to hiding soils from the manufacturing environment, but also allowed me to maintain a power distance with the employees and managers. The saying that "clothing makes the man" (or woman!) is still relevant, but perhaps with cultural bearings in mind.

One other particular element of attire must be cautioned and that if for gender. In the US, wearing a suit might be acceptable for a woman and wearing an earring might be acceptable for a man. Not all countries, territories or even companies are so free with these practices. If a woman is to make a trip into the Middle East it would be best to wear the most modest clothing possible. Expectations of modesty vary dramatically in both social and business environments.

Err on the side of caution (Figure 4.1). When in doubt I wear my finest suits and ties to impress and gain respect. It is always easier to remove the coat and tie if I'm over-dressed. As a final note, those individuals in the London course were wearing less formal attire by the third day of the class. Even I removed the tie, but kept the jacket, which indicated the comfort that individuals had within the group. Even if it seems trivial, be aware of what you wear. Of course, if you are not visiting the country or participating in video conferencing, this topic will be of lesser importance, but will come up some day as business becomes more and more global.

Type	Appropriate Attire	Inappropriate Attire
Business Dress (MEN)	- Conservatively cut business suit in navy, black or gray. Subtle pinstripes or plaid fabrics are acceptable. (**Required for presentations**) - Long-sleeved shirt in conservative, coordinating color. - Conservatively patterned tie in coordinating colors. - In less formal situations (e.g., Boeing visit, Dean's Global Business Round table), slacks, sport coat, and tie are acceptable. - Dress shoes with dark socks.	- Casual or business casual attire. - Bright or neon colors. - Casual loafers, boots, suede shoes, or any other casual style of footwear. - Hats or caps of any kind.
Business Dress (WOMEN)	- Conservative business suit or tailored dress with jacket (**required for presentations**) - In less formal situations (e.g. Boeing visit, Dean's Global Business Round Table), a professionally cut pantsuit or slacks and a blazer. - Hose and low- to mid-heeled pumps or dress shoes. - Makeup and appropriate accessories.	- Casual or business casual attire. - Sheer fabrics, neon colors, very short skirts (more than 3″ above the knee), tight blouses, low necklines, any kind of glitter. - Open-toed shoes. - Evening makeup or jewelry. - Hats or caps of any kind.

Figure 4.1. Proper Business Attire [2]

One exception to this would be a time when in China, I had been at the same factory for three weeks. At the end of the trip, I told the president that I would play basketball with the plant teams if they wished. I'm not too tall, but back then, in my late 30s, still wasn't too bad. The employees thought it was great that an American "business" person would play with them. I wore shorts and one of the team's T-shirts. They all wanted photos and was quite the "celebrity" since nobody from out of the country had ever done that. This time around, it wasn't to gain respect, but to be apart of the team. They were thrilled!

Finally, a very eclectic scenario on my first trip to the United Arab Emirates. While waiting in the hotel lobby, there was a group of Japanese business men in suits having a discussion. Up on the second floor balcony, there was a woman playing sitar and another belly dancing. At the front desk there were men checking in guests in the traditional Kandura garb. Then, through the door walked a very tall blonde family, parents, son and daughter, probably all six feet or more. The young daughter, probably about 15 or 16 was wearing VERY short, shorts. My first thought was "you are in the middle east young lady. What are you and your parents thinking?". An associate of mine explained it this way. The United States was called the melting pot where traditions blend together, while the UAE is considered the jelly-bean bowl. All individuals can be kept together, but retain their identity. Since then, I've observed local women fully-covered in the swimming pool while foreigners are in bikinis. Surprising to see that tolerance, but different than other places in the middle east. Respect is key.

Politics

Another area a project manager must place some focus is the politics within the project work areas. If a country has little affinity to the same cultural beliefs, it is quite likely that the government operates in a different manner. Communist China and monarchal Monaco contrast deeply with the work of a democratic US. If you are traveling to the country, it is wise to know the government structure and level of commitment to foreign investors and such.

An experience that an associate of mine made on a trip to Thailand a few years ago was interesting. In September 2006, Lieutenant General Sonthi Boonyaratglin launched a successful overthrow of the government while the prime minister was in the US for a UN meeting. There was little violence, and only for the resistant party being overthrown. In a few days the new military government had established their clear leadership and things seemed settled from an outsiders perspective, but this again changed in December of 2007 when democratic rule reestablished the civilian government. What seemed very interesting is that, as the story was related, the monarchy is so well respected that if the throne had disagreed with Boonyaratglin, that the people would have forced him out. Inter-political challenges could have threatened the safety of foreign workers, but in this case, it had no effect. [3]

Another relevant challenge in politics is the work of contractors in Afghanistan and Iraq. People place their life on the line knowing that when they take the flight into these countries they may not return alive. The odds of an actual altercation is not extremely high, but this is due to the training most contractors receive on how to protect themselves. Should there be another turn of events in the ongoing battles, those working in these countries face serious risks.

Not every international work situation has this level of threat to the individual or to the project. Usually the threats of international politics will affect the scope, schedule and budget more dramatically than anything else involved. In the grand scheme of the project, planning for the various scenarios that might occur will help to prevent confusion over permitted projects (scope), the work ethic (schedule) and the monetary stability of all nations involved (budget). Planning for necessary approvals from government agencies and having a plan to share the project with authorities in a positive light will help reduce anxiety and risks.

Another area of politics is internal politics in an organization. While employees may be driven to work for the company in some cultures such as Japan, others are driven to work for themselves such as the US. Change will be difficult in any organization because of the maxim that people are creatures of habit. "In universalist cultures, leaders should expect employees to grumble about things, but broadly fall in line. In relationship-driven cultures, on the other hand, unwelcome changes can be met by open rebellion, as old loyalties count for more." [4]

Prepare yourself for internal and external strife.

Ethics

Another area which seems to be a tremendous challenge is that of ethics. This area of discussion is based on high US standards which are governed by US Professional Engineer's code [5] of conduct, the PMI Body of Knowledge [6]: Ethics and national, state and local rules which permit specific levels of contributions, information sharing and contact. Beyond the laws and rules of conduct which are established, companies may have their own guidelines or rules which establish strict boundaries.

When dealing with individuals in other nations, there are various sets of standards that might be considered acceptable and they cannot understand the limits which you are setting. In western nations there is a similar code of conduct, but elsewhere, the rules change. Let's look at some examples:

In Russia, a lot of black market equipment is sold. Back in the early 1990s a young teenager I knew went on a school trip and purchased night vision goggles that were only one generation less than what the US military just received and got it for less than 10% of what similar US "private-market" materials were being sold for. One might wonder how this happens so easily, but it was not considered that unusual for the individuals in that area. While he visited he said that every day he could find another shop with what he assumed were stolen goods including weapons. Frightening to think about how accessible this material was to a foreign-speaking tourist, who was under age at that! The individuals selling it didn't seem to mind at all.

There was a Chinese company that made a manufacturing error, which created approximately $250,000US in equipment. Based on the receipt of the first container of equipment, the materials of the next seven containers would need to be rejected as well. The error was of a nature that a potential fire hazard existed and the machines required rework or would be rejected. The information was passed to the Chinese manufacturing engineers who thought about it and proposed a few ideas, but none would pass stringent US fire codes. One of the American engineers proposed a fix that would likely cost $30,000US. That would be a lot better than $250,000 in wasted equipment plus shipping costs. The Chinese manufacturing manager was not inclined to eat $30,000 and instead sent an email to my personal, private email rather than my professional email address for the organization. In the email he noted that as the project manager "you are the king of this project and can make this go away. If you approve the equipment, they will accept it and when I visit the US next, I will show you my personal appreciation." My immediate response was to reply with a cease and desist and a strong piece of advice that they never send something like this again to anyone in this organization or to any of the companies with which they work. It is just plain wrong. After a follow up phone call to reiterate the advice they seemed to get the message that we find that behavior unacceptable. This email and the conversation were also shown and described to several individuals in the company to assure them that no such bribe would ever be considered. It is doubly concerning that it was a safety issue and everyone appreciated my candor and diligence with which the manufacturing contractor was reprimanded. Shortly thereafter, the US company began searching for alternate sources in China that respected US ethics and would not willingly compromise safety.

Another example of ethical challenges would include gifts. Some countries, particularly Japan and Korea, have tradition in awarding the leaders of projects with large sums of money or extravagant gifts. While the intentions are well meant, the amount of money or gift is frequently beyond permitted contracts or obligations to organizations. Many companies limit such gifts to $100 or even as low as $25. Some companies outright forbid gifts of any sort to prevent confusion on the topic. In one circumstance the project manager was given a gift of a new car in front of the sponsoring company's employees. The project manager was shocked and in front of the client and their employees did not want to create a scene or offend them. Of course, future work depended on keeping the client happy. To this end the project manager would have to strategically smile for the employees and decline the gift. One thought is to shake the hand of the executive offering the gift and note how nice the gift is in a loud voice, but lean into the executive and state "This gift is great, but I cannot accept it. Let's discuss later how we might share appreciation. I already have an idea!"

Personally, when such a large gift is made, my inclination is to ask that a contribution be made to a non-profit organization for education or children in need. Make certain that the donation is made in the name of both companies. It generates good will, brand recognition for both parties and will get you out of a sticky situation usually in excellent shape to do business together again.

Knowing these stories and how to successfully navigate the international understanding of ethics is tricky. Remember though, always take the high road and you will succeed in the long-run in international projects.

Etiquette

The fact that rules of dress are changing as noted in the prior section would indicate that perhaps other areas of work life are becoming more casual as well. This isn't always a good or bad thing, just another component of the big picture to consider. As American culture has stopped wearing certain pieces of clothing at work, so too the culture has terminated the use of "Mr. and Mrs." for most coworkers, subordinates or superiors in the organization. With this change, it takes away the power distance [7] of position and is intended to make all employees approachable. This is not the same throughout the entire world. Those in London may still use the title of the individual at least during the first meeting to show respect, particularly if they are a high-level executive or an important client. Those in Korea, Japan and Thailand have traditionally used the title and first name to be respectful, but still personable. Be aware of your place within the business structure and situation so as not to create an underlying feeling of discomfort for the team.

Many other forms of etiquette remain in place for American work places, but this doesn't mean that the same traditions exist everywhere. Consider that in Japanese culture, to this day, one does not interrupt or even speak when the boss is present unless you are directed to do so. What seemed very unique in one meeting was that the manager of a department came with two individuals from his group and they sat on one side of the table while the US employees of the other company, manager and two employees, sat on the other side of the table. It might seem the adversarial seating arrangement, but it was traditional and respectful. During the meeting, individuals from the US side all spoke regarding various topics and chimed in wherever they felt necessary. However, on the Japanese side of the table, the manager did all of the speaking. The other two employees would occasionally lean over to whisper something in his ear and then it would be passed on to the other company. It seemed quite unusual and less productive to have a meeting where some of the employees could not speak up directly. If one is to understand the cultural background, it becomes obvious that this is a matter of meeting etiquette.

Hopefully as a project manager it is obvious not to take the seat at the end of the table unless your are directed to do so by your host or it is your facility and your seat to lead. On many occasions the project manager is not the "head" of all parties in the room and that seat should be reserved for a client or other high-end stakeholder.

Taking calls or sending email in the middle of a meeting might be commonplace for some organizations that find the flexibility and "open meetings" productive, but this is not always the case. One of my lines prior to starting a meeting is that "We all are busy and have professional careers to support and it is understandable if a call is necessary, but please leave the room. The only reason for your cell phone to ring during this meeting is that it is your birthday, so if one goes off, everyone in the room is to sing happy birthday. If it is not your birthday or you do not wish to have it sung to you during our course, please set your phones to vibrate, silent, shock or stun!" That usually gets a good laugh and the class begins rolling. Usually individuals are sensitive to those around them and very rarely has a phone actually rung during a class and one group did sing! Take into account whether the meetings you participate in are open or closed meetings and if you are leading the meeting, instruct those present so that expectations are clear.

Consider these cross-cultural cues to achieve success [8]:
1. Develop a detailed understanding of the environment.
2. Do not stereotype.
3. Be genuinely interested in cultural differences.
4. Do not assume there is one way – yours – to communicate.
5. Listen actively and empathetically.

Equal Opportunity
Project managers have to find the best talent to accomplish the project scope on time and on budget. This is hard enough to do when using an open talent pool, but when situations are restricted due to

bias or cultural tendencies, it will make it even tougher on the project manager.

Some countries such as the US have equal opportunity clauses or affirmative action laws that require companies to hire based on skills and not on gender, race, religion or other criteria that might discriminate. This is not always that case throughout the global competitive markets. While western countries continue to make slow strides in this direction, other countries have not made such commitments or have actively avoided doing so based on cultural beliefs, national religious preferences or simply based on dominance of one group over others. This is where challenges might begin to creep into the situation.

Where possible, the project manager should fight for the employees that have the greatest benefit to the project, regardless of the aforementioned discriminatory divisions. If it is possible to hire a qualified woman to do the job, then consider doing so. Of course, there may be challenges in this. Even if the project team is open and everyone is satisfied with the instructions from her, there may be more challenges from external stakeholders such as vendors or government officials that feel they can take advantage of the situation. As the project manager, your job is to make sure that the project is completed and the staff on the project are well-supported. This may mean playing referee once in a while for someone being unfair in situations. Only ten years ago, in the US, a manager in a department told me that a girl could never do the job in the R&D facility because it was too physical. After closing the door and giving him a lesson on equal opportunity, I also stated that it only seemed this way to him because he had hired mindless thugs to work in the group. I filled the two open internships, the jobs with the most physical work, with young ladies and they did quite well, learning to use brain over brawn (and machines over muscle)!

Sometimes the project manager will have to intercede on behalf of the suppressed minority. In fact, for many American project managers, they are apart of one of these minorities and can certainly relate to the challenges for project team members. Actively encouraging and supporting staff will not only win over employee favor, but will help you develop negotiating skills as well. This is less problematic in 2020 than it was in the 90s or before.

Dinners

Business settings shift from the office to off-site dinners on frequent occasions and for some rookies in this situation the temptation is to believe that business is done for the day. Perhaps the jackets and ties come off from a formal setting or perhaps the mood is a bit lighter, but the assumption that all caution can be thrown to the wind would be foolish. Individuals from the executives to the entry-level staff are being judged on character, conversation and of course business etiquette.

Looking at the situation from this perspective, one should ponder whether they will be taking a lead role or an observation role at the dinner as well. Certainly dinner can lead to interesting personal conversations and life stories that are more appropriate in this setting than the office, but make sure that if the conversation were recorded in a set of minutes for the dinner meeting that you would be comfortable having them read back afterward!

Two ladies, Pamela Holland and Marjorie Brody wrote a book called Help! Was That a Career Limiting Move [9] which highlights poor behaviors that are rampant in business. Some individuals don't realize it because they work alone frequently or just never paid attention to their manners. This is a recommended reading in the final chapter.

General dinner manners such as keeping your elbows off the table, chew with your mouth closed and be careful how much you drink, if it is permissible. Consider the ramifications of a first meeting and you begin by sitting slouched at the table, end up spitting food on your client and then after a few too many drinks make an inappropriate comment. All of the work performed during a successful day of meetings will be lost at the dinner. If there are further meetings the next day, recovery from such a blunder might be possible, but if not, the client may have enough time to consider other options to using you and your firm.

Email, Phone and Communication

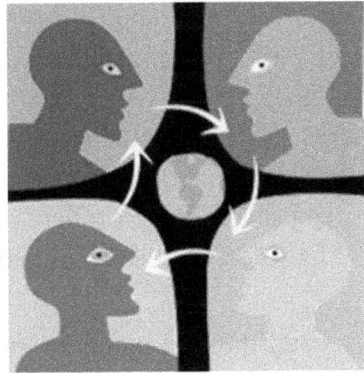

Communication will be repeated again in this book since 90% of a project manager's job is communication [10], but specifically for international efforts, the choice of communication method is critical. Any project manager should begin a new project of significant size with a kick-off meeting that introduces all of the players to one another in person. It is the healthiest way to make everyone comfortable with who will be involved throughout the duration of the project and get a sense of the type of people, roles, challenges and needs that will arise. If one does not give a serious consideration to this first meeting, don't expect stellar performance in the project.

Realize the following communication percentages by mode [11]:

7% of communication comes from the words
38% of communication comes from the tone and inflections
55% of communication comes from body language

Without a phone you are losing 93% of the message on the receiving end! Not a good chance of making sure the situation is clear for everyone. While this is an old study, it's clear that humans are visual and audio-sensitive creatures.

If there is absolutely no way to get everyone together in the same room for a meeting, then the next best effort would be to connect via internet each of the locations with video-conferencing and make everyone feel as though they are in the same room. One method that is beneficial is a war-room with enough connections to allow each individual to chat over the internet during the meeting. They can see one another over the video conference in the other room and then send messages. This is a technical challenge that should be fairly simple with today's networking capabilities, email, instant messaging, chat, etc. Some of the netiquette basics would be to keep your camera on, stay focused, use the mute when appropriate, try to keep pets off camera where possible and make sure you're not doing "other things". This is quite the temptation with the camera off.

Beyond the first meeting, individuals need to feel that they have a connection with the parties involved. The judgement on who should be in meetings, how frequently to meet and how to run the meetings is up to the project manager based on experience and the particular demands of the project. A good heuristic or rule of thumb would be to call a meeting at just the time when everyone needs to exchange crucial information and make decisions on that information. The discussions in the meeting can happen more effectively and efficiently than through a series of disjointed calls and emails.

An error made a few years ago was that one client wanted all communication after the first trip to be done by email. In this manner the accent of the foreign workers would not challenge one another and the expense of phone calls would be eliminated. At least that was their logic behind the decision. As noted above, the problem was that the emails only captured a portion of the intent. Communications by email get 7% of the message across. Thus email has limited benefit unless purely communicating data and not ideas. When a weekly conference call was added, the performance of both the local and international teams improved. Individuals had to repeat exactly what they understood were the goals for the next week and frequently the misunderstandings could be clarified.

To the best of your ability select the proper mode or method for communication, the frequency, the parties which need the information and control this flow of information so that individuals are not left out that need details, but that proprietary or confidential information is not leaked. Both are critical to success in today's business environment.

Chapter 5
Exploring Romania, China, India and USA Projects

"160 cars can drive side by side on the Monumental Axis in Brazil, the world's widest road" [1] (It's a myth, but it sounds good for tourism)

Now we will take a look at some comparisons of various countries and the relative benefits and challenges of each within the scope of project management. Note that these countries are four of the five largest countries by population in the world and each holds certain benefits for discussion of global projects. We could take a number of European, African, Middle Eastern or other Asian countries into the picture as well, but these were selected for sake of comparison.

There are a few more topics to discuss related to differences in cultures that we wish to share and then some charts should help with the comparisons.

Power Distance
One phrase that has arisen a few times in prior chapters was that of power distance. This is a term used to describe the amount of power that individual employees have when working directly with their manager in various circumstances. [2] If one has high power distance from their manager, they are not permitted to discuss certain topics in public and may even be prohibited from

breaching the subject in private. High power distance means that power comes from the title or position held within the organization.

Conversely, societies with low power distance would permit open discussion regardless of rank and order within the company.

This topic was specifically related to the difference in Japanese culture where the boss speaks on behalf of everyone in the group as opposed to American business where everyone has an equal chance to speak. Of course, this does not mean that everyone gets the same opportunity to make decisions, but rather to voice their opinions. It is quite conceivable in the US for a low-level manager to get the chance to speak directly with vice presidents or even the CEO of the organization. It may not happen every day, but it usually considered fair practice.

If you participate in meetings or discussions in foreign nations, be prepared to open the avenues for project team members to speak openly. They may be reluctant to do so if past culture would preclude this openness. You may even need to directly speak with them in the group and ask for their opinion. If they still shy away from speaking openly, you may need to have a private conversation encouraging them to open up. Even after this encouragement, their inward personality may require follow-up to make sure that they have shared with you thoughts that might not have been presented to the group.

Don't overlook the personalities and culture that might hamstring your efforts.

Individualism vs. Collectivism

Another area of culture that can be company-driven as much as country-driven would be that of the affinity toward collectivism or individualism. The focus here is on the attention given to individual or group success. This depends on a number of factors, but an obvious one is compensation and reward for work well done. In an individualistic society, reward is based on the merits and success of each individual employee contributing to the effort. This is a very capitalistic view, but this is how the US operates. Each employee has their own negotiated salary, benefits and reward plan. It lends to everyone striving for their own best interests, but as long as the compensation works out to the best interest of the company, that is determined acceptable.

However, other societies focus on the good of the group or company. Everyone focuses on a "we" approach and everyone is rewarded if the team succeeds and nobody is rewarded if the group fails. This lends to a team approach where everyone helps out everyone else to achieve all goals. However, if somebody strong continues to support the team, they receive no more reward. If there is somebody weak and he or she is continuously carried by the group they may learn to float. This typically is not the case in collectivist societies because one grows up with a tendency of helping others regardless of the reward structure.

We have focused on the countries, but pay attention to the fact that companies will encourage or discourage one behavior or the other as well.

Uncertainty Avoidance

Another tendency is to separate the level of risk acceptance or uncertainty avoidance from one culture to another. Those that see a situation and avoid risk will generally be safer, but less well-off. This is a general statement of risk-reward scenarios. In the US, most individuals seek the "American Dream" which is to have a secure income and own a home. This requires a lot of effort, but with some risks taken as well, almost anyone can open their own business and become moderately wealthy. Or lose everything. The expectation is that with a little money and the willingness to risk that money, you can multiply it through effort and business focus.

Some societies do not encourage this behavior and are far more satisfied to accept lower incomes and fewer amenities in life with the feeling that risk should be averted and accept one's lot in life. Such aversions may come from governmental rule (communism and dictatorships), religions (accept your lot in life) or tradition (it's always been this way). Whatever the case, it does not make these cultures wrong, since there is some security. The likelihood of absolute bankruptcy in the US is quite high due to individuals overestimating their own capabilities and losing everything. Between July 1, 2007 and June 30, 2008 nearly a million individuals or businesses filed for bankruptcy in the US. [3] During the 2020 COVID situation, it was anticipated that as many as 40% of companies could not handle 4 weeks without income.

When handling projects in other cultures, attentiveness to the risk behaviors will help secure lower cost labor or better deals, but might also lead to risky contracts. Determine the level of risk acceptance by the stakeholders in your organization and pass these beliefs down to the project team members. If it is high-risk acceptance, challenge those in the organization to think of new ideas to achieve new levels of success. If risk is a high threat, then enforce strict reviews of all project work to eliminate this risk and enforce this as the project culture as well.

Cultural Adaptability or Ethnocentricity

The definition of ethnocentricity is:

Ethnocentricity: The tendency to believe that one's cultural values and ways of doing things are superior to all others

Another area of concern should be the attachment to the culture's own beliefs. Some societies and cultures are very open to other practices and beliefs, while some are very closed off and will be offended should you make any attempt to introduce them within the country. The United States has been called the great melting pot [4] for many decades, and yet some would contest that the pot didn't melt, but simply mixes. Many individuals maintain their heritage and practices when they come to the US, but the acceptance of other cultures, religions and general diversity is embraced as a good thing. Some countries are resistant to this practice and caution to abide by existing practices must be observed.

Can anybody imagine a Middle-Eastern woman roller-blading down the streets of Baghdad in a bikini? This may be a striking example to select. Anybody in their right mind would obviously not attempt this. Yet, every day people of all races and backgrounds are free to do so on the beaches in California and Florida. Culture makes a difference.

Within project teams, individuals may have issues dealing with somebody of conflicting race, religion, gender, dress or other definitive difference and the project manager needs to help smooth these differences and help team members to work through the differences and perhaps see the benefits of both methods. A successful project team does not need to change the beliefs or all practices of the team members, they must learn to work with the tools they have.

	Brazil	China	India	USA
Political stability	3	4	3	4
Worker skill	3	2	2	5
Worker supply	4	5	5	3
Culture compatibility	3	3	2	N/A
Infrastructure	3	4	3	5
Government support	3	5	4	5
Product-to-market advantage	4	3	2	5
Transportation	3	4	3	5
Educated workforce	3	3	3	5
Utilities	4	4	4	5
Telecommunications	4	4	5	5
Vendor supplies	3	5	3	4

Score legend
5 = excellent
4 = good
3 = acceptable
2 = marginal
1 = poor

Figure 5.1 Assessment Matrix Project Site Selection
(Manufacturing)

	Brazil	China	India	USA
Diversity	2	2	2	5
Etiquette	3	2	2	3
Dresscode	3	3	3	3
Language	2	3	4	5
Time sensitivity	2	3	4	5
Currency/Inflation	3	3	3	3
Food	4	4	3	5
Security	2	3	3	4
Religion	4	2	3	5
Holidays (week or more)	Feb 21-25 (+ week ends Carnival)	Jan Feb New Year October festival	Only single days off through year	Last two weeks of December
Family Life	4	4	4	3
Ethics	2	2	2	4

Score legend
5 = excellent
4 = good
3 = acceptable
2 = marginal
1 = poor

Figure 5.2 Assessment Matrix - Project Site Compatibility (Social)

The point of these charts is to assist in selecting compatible work situations, if the project permits such selection of team nations. However, in most cases, the project manager is dictated where the project will be located and thus the country selection may not be relevant to them. This type of exercise though should be done as a project team to help identify the differences among nations and therefore raise awareness to the unique situations that will be faced. To illustrate the differences, it might seem appropriate to share a few stories here about each of the nations and how an outsider from the US might be struck by the differences.

Cultural Issue	Variations		
Relationship to nature	Domination	Harmony	Subjugation
Time orientation	Past	Present	Future
Activity orientation	Being	Doing	Controlling
Nature of people	Good	Evil	Mixed
Relationships among people	Individualist	Group	Heirarchal

Note: The line indicates where the United State stends to fall along the se issues

Figure 5.3 Kluckhohn-Strodtbeck's Cross Cultural Framework [5]

Relation to nature – How people relate to the natural world around them and to the supernatural

Time orientation – The culture focus on past, present, or future

Activity orientation – How to live: "being" or living in the moment, doing, or controlling

Basic nature of people – Whether people are viewed as good, evil or some mix of these two

Relationships among people – The degree of responsibility one has for others

Romanian Construction

In the US, if you wish to build a home, office building or other structure, you must have suitable financing to complete the project whether it is cash-funded or through a secured loan. In this manner, even if the building is not rented or occupied at completion, the facility is still finished. The US has codes and regulations in the cities and suburbs that require a building to achieve a certain level of completion to prevent storm damage, discourage vandalism or increase safety where children may be tempted to play when the site is unattended.

If you ever travel down one of the roads in Romania, you might pass a number of homes, office buildings and other structures that are only partially completed. While the intent was to build the full structure for the intended purpose, many remain in various stages of completion for years. The Romanian government does not regulate whether the buildings are finished and thus completion is optional! It would seem a bit counter-intuitive to do this, but the thought process is to complete as much of the project as possible and then get more funding to do the next stage.

This same process of "spend what you have to get the project going and then work to get the rest of the money while the building sits unfinished" is not only in Romania. With inflation considered for some currencies it is reasonable to put your money into finished products and construction labor prior to any further devaluation. Thus, the building of structures without full funding isn't necessarily wrong, but requires adaptation to the economic demands and cultural boundaries of acceptance.

Chinese Manufacturing Sites

Upon a visit to China a few years back it struck me that there are several significant differences at Chinese manufacturing sites which I had not expected. For one, in the US a manufacturing site is usually built with serious power upfront and the facility is designed specifically for the shop that is going to occupy it. In many circumstances this is not true, but for the most part, companies have special design needs that they include into the original plans. In China, particularly the Shanghai and Guangzhou areas, the facilities are built of concrete with large windows and moderate power to the building. Once the building is occupied, the company changes whatever seems necessary internally.

After a brief tour of one facility, I used the washroom for the workers and found that three of the four urinals were leaking to the floor (water only!) and that the installation of the plumbing had been made crooked. With a few minutes repair and a couple of proper compression nuts the leaks could be repair and the pipes could be straightened, with the fixture mounted an inch higher. Upon joining the managers of the facility again I asked how old the facility was and who installed their plumbing. The facility managers laughed and told me that the building was less than three months old and it wasn't a problem for the plumbing to leak. They were used to it and the building was only expected to last for five years!

Don't take this as a derogatory remark against all manufacturing sites in China, but rather one facility encountered. Others are quite nice facilities, but it depends on the grade of expectation for your organization. The Olympics in Beijing were the greatest entertainment spectacle the world may ever have witnessed to that point, but the grade or level of product or service expected can be quite varied.

One thing that was beneficial at this facility was the window-based lighting which permitted the workers to operate under sunlight conditions during much of the day. The challenge was on a particularly cloudy day that the workers could not see paint defects. The American quality manager visiting told the plant managers to turn on the lights so that they could work properly. After leaving the facility and getting in the car, the lights inside were already turned off! The plant managers were avoiding cost at the sake of worker comfort or industrial engineering benefit [6]. They could not be convinced of the benefits of paying for the lighting on cloudy days.

Another interesting aspect of the Chinese manufacturing world is used parts. At one facility, automotive parts were being refurbished or remanufactured for use in China. Considering that the metal itself was in reasonable shape, the primary focus was cleaning the parts. However, we didn't get to see the cleaning operation at first, only in the inbound and outbound parts. Upon request the party visited the cleaning area where we expected to see large washing equipment with either water- or solvent-based cleaners. Instead we found two dozen women with red shop rags rubbing the parts. No fluid at all. Just elbow grease. Considering my engineering background is in industrial equipment manufacturing and design for the parts washing industry, this was absolutely shocking. Why might they do this? At that time the economics of Chinese labor were so much cheaper than equipment purchase, repair, and chemistry handling that shop rags and elbow grease were the best option.

Also, the manufacturing facilities were based heavily on manual labor. This was no surprise and is a primary reason for purchasing products from China. However, the sheer magnitude witnessed this is a different story. My brother works in a machining shop in the US and he operates four CNC machines simultaneously to keep parts rolling. The main shop floor is occupied by perhaps a dozen operators and fifty machines. Entering the Chinese facility of perhaps double the size demonstrated about 100 workers scampering about, each handling a different task, but with several hundred end products, sitting on the floor! There was no production line. There were no power tools. There were no quality inspection tools. The facility was filled with product and individuals randomly selecting a machine they could get to for the next task. Understand that the primary reason for the US party to visit this facility was to improve production quality and therefore these initial observations within the first two minutes made apparent the drastic changes that were necessary. Total costs for improvement were about $5,000US and the operation proceeded to improve from 50% out-of-box issues for the first container load to less than 2% by the 8th container. After three years of operation the expected failures are still under 2% with on-going monitoring.

The Chinese workers at the lowest cost facilities are based in a manual mode and thus, to improve performance, some encouragement might be necessary. A couple of terrific benefits that these workers offer is willingness to learn how to do something differently and diligence. They will repeat the task until it is right. We spent time until 10pm or 11pm one evening working with the same individuals until we were both satisfied it was right. The workers didn't even question leaving at 5pm, but continued on. A tremendous asset in facilities that need to learn improved methods.

India and Tech Support Centers

A few years back one of the clients for whom I worked had decided to move their entire technical support operation to India including internal IT support for several thousand employees. The following story is not typical, but quite humorous if you can imagine a 21-year old American designer trying to get help from a new tech support individual in India:

The American designer was sitting in a cubicle within ear shot and was working on a particular 3D model which was not blending two spherical radii the way intended. However, each time he made a change, the computer crashed and upon reboot did not accept the current password. A fairly complex issue when software and platform are interfering within password protection. In any case, the designer called the 800-number and put his phone on speaker. After quickly relating what happened, the support tech stated that he was glad to help.

The tech support rep then proceeded to ask whether the computer was on. The designer stated sarcastically, "Yes, I just explained what all happened" and went on to reiterate the situation again, but a bit more slowly. The tech support said "thank you. Is your monitor turned on?" Again the designer sarcastically reminded the tech support rep that he had gone through the entire scenario twice and he reiterated a third time the issue. The tech support rep followed up with "good, now I understand. Is your printer hooked up to the computer." At this point the designer got frustrated and asked for the tech support reps name. "My name is Chuck" was the answer. The designer, absolutely furious with the lack of actual support shouted at his speaker phone "Listen, we both know your name isn't Chuck! Now get me somebody who knows what they're doing on the phone!" Two points to be made here. Firstly, the American designer had a slightly slurred Midwest accent, but rarely got upset at anything. Secondly, he did ultimately demonstrate the demanding behavior some Americans have. Customer support is not a luxury, it is expected. However, to "Chuck's" credit, he remained calm, despite having a total lack of experience and an impatient American on the phone with him.

At first, the cost reduction appeared to be well worth the cultural and language differences. However, there were several incidents similar to the one noted above and the organization brought the IT support function back to the US and since has heard few complaints. The issue with the support from India was that the employees had a disparate knowledge of English and limited understanding of US urgency to get to the point quickly. The patience with which all of the Indian customer service reps answered was impressive though.

These are only a few of the more dramatic experiences that can be shared. Many positive and less colorful experiences have taught me to be open and ready for cultural differences in each country and within larger countries.

Chapter 6
International PM Initiation and Planning Tools

Any fool can make a specification complicated. It takes skill to make it simple. The Soldier's Rule. [1]

The quote that opens this chapter is something that I have used for many years as a motivational tool to keep all work that I perform simple. The acronym KISS for keep-it-simple-stupid is another way of saying that less complicated devices, plans, and projects are more likely to succeed. Another quote that I have always liked was by Einstein "Make things as simple as possible, but no simpler." [2] In the interest of leaving past behaviors behind, individuals must challenge themselves to remove any task or specific responsibility from the project if it does not serve a useful purpose or elimination would have limited detrimental impact on the outcome. In these cases, there could be a variety of reasons why the unnecessary task remained. Many of times the root is a behavior, habit or tradition that also must be far more diligently weeded out. Any time somebody states "because that's the way we've always done it" a caution flag should go up to indicate that thorough review is in order.

Projects are to be planned out thoroughly so that "the work required, and only the work required, to complete the scope" [3] is included in the project plan. Hence the initiation and planning phases are critical to the success of the project. If the project begins with a faulty plan and is perfectly executed, controlled and closed, the project will still be faulty!

In order to evaluate a proper initiation of a project, the project team leader must be selected for the project based on past experience, relevant technical knowledge and, particularly since this book discusses international projects, familiarity with the countries and peoples involved in the development of the project. We will address the traits of the project managers and specifically international project managers in a following chapter. The success of a project depends on many factors, but the capabilities and tools, which a project manager will employ is a large portion of the battle. The project team is another angle to consider and we will again address the rest of the team and selection under the chapter dedicated to Individuals Suited to International Project Management.

At this juncture, we will focus on specific tools and techniques that are used at the front-end of a project to make sure that the targeted goal and expected outcomes are achieved. We will look first at some basic project management techniques as a basis and then address specific international items that will require attention.

PMO or Templates

As soon as the project manager is identified, efforts should be made to define the project team members that will be necessary for the duration of the project. This is the core team that will support the project and project manager for each element of the project and therefore are long-term members. Individuals that support for a limited time might be SMEs (subject matter experts) and are not necessarily required in the initiation phase if they are truly consulting for the project and may only be necessary for one portion of the project.

Once the team is locked in, or at least mostly defined, the team should seek out a meeting with the project management office, known as a PMO. Large companies tend to have a PMO in order to keep the project managers within the company on the right track. This might mean helping with training in areas for which the team or one participant is not prepared. The PMO therefore is a training group (supportive PMO). The second purpose of the PMO is to provide templates and consistent documentation requirements for project management, whether they be domestic or international in nature. Of course, projects are unique, one-time endeavors, so some modifications to the templates may be permissible within limits. The PMO might also help the team review the *product* scope and look for areas of clarification that will be necessary in generating the *project* scope. There are also controlling PMOs which means that they have authority to "mandate" guidelines and processes. The PMO with most authority is a Directive PMO which actually takes responsibility for the projects. The Director employs PMs who run the projects.

If the company does not have a PMO or director of projects, the project manager might be left to their own devices. In too many circumstances, there are project managers that look to overthrow past experience and feel that they can do everything better. This may or may not be true, but it is completely unknown, if the templates, tools, methods and past experiences are not tapped first. To this end, even if there is no project management office available, a good project manager will seek out the experience of past project managers that have worked on similar projects or at the very least may have templates to use. The best place to find planning tools for a new project is based on historical information. Documentation is in place for a myriad of reasons, but one of them is to learn and improve upon for future efforts. Use them!

Stakeholder Analysis

Once the documentation and past work is reviewed, the team is educated from past successes, failures and mediocre performances. The next step should be a thorough stakeholder analysis. This is a basic requirement for any project regardless of size, location or product/service being prepared.

Various methods exist to develop a suitable list of stakeholders, but using the information from similar completed projects is the logical place to start. That will give a healthy start and reduce the amount of work in researching individuals, preferences and specific needs. Of course, filtering the list for those stakeholders that are not relevant would be logical. Also, the stakeholder may have changing needs over time. If the project is a few years old, conceivably the same individual will have different needs and wants. Ten years ago a portable CD player was the "in-thing" but today, they cannot be given away in light of all-digital tools in cell phones having taken over the market. Today's markets and economies move faster than ever and with each passing year will continue to accelerate. Review what the stakeholders need now.

The next step in the stakeholder process would be a brainstorming session where the ultimate expectations of the project are defined by a basic product definition. Once the product is defined, the project team works at different angles of who would want the product. Who might be affected by the product or service in a positive manner? How might the product or service negatively impact external parties based on noise or air pollution? What regulatory bodies might be interested in the outcome of the project? If you were a "nobody" to the project, how might you want to get involved or wish to stay ambivalent? Through an open discussion of this nature, interesting outcomes will clarify the need to define the project scope more thoroughly.

Never assume that the stakeholder analysis is complete. Throughout long construction projects, new stakeholders may move into the area or new regulations or regulatory bodies might become interested. As the stakeholders change or their needs and wants change, update this list. It will be crucial to project buy-in as activities progress. Planning doesn't stop until the project is nearly over.

Scope Clarification

One area of confusion for many engineers is the term scope. Usually a scope of work identifies the work to produce the end product that will be delivered. Sometimes this is a set of drawings. On other occasions it would be delivering a batch of component parts. Yet further, another scope might require delivery of the entire end product or service. Regardless, keep in mind the general differencefor product definitions (requirements) and project definitions (deliverables). Sure, the two go hand-in-hand, but keeping it clear that a project manager focuses on the project's scope (what), schedule (when) and budget (how much), more than the actual product.

At the initiation of the project, generally the project sponsor will hand the project manager a list of features and attributes that the product or service must provide to the end user as part of the project charter. During the project team review of past projects and stakeholder analysis, usually there are some questions related to the scope of the *product* or *project*. If these are feature and attribute-based, it is a good time to clarify with stakeholders what might be expected and gain necessary research with the support of the project sponsor. If the product is ill-defined, and the project team delivers what the sponsor asked for without sufficient research, ultimately the project will have been for naught. Polite challenging of the product features, within permissible limits, will help ensure success during the early

89

phases. Asking 1000 questions might take a couple days up front, but it might save months of rework on the back end.

Also, as noted above, the project should clearly identify which of the triple constraint goals is primary, secondary and tertiary. If the scope must be met without question and the schedule is extremely important, then the individuals footing the bill, must be made aware that costs will be elevated. If the scope is critical and the budget is limited, then schedules will have to be spread out to achieve success. The project sponsor, stakeholders and individuals may not be very willing to accept this, but if you pull out the triple constraint and explain the ramifications, generally they understand and grudgingly will organize the priorities.

Upon completion of the investigation phases, the scope can be clarified and signed off upon by each of the required parties. This will be the core document around which all work will be based. Make sure that this is right. If a delay has to be made in the project for a few days to lock in the scope, to repeat, it may save several weeks or months of rework at the end. Be patient and persistent to ensure that all parties domestic and international expect the same outcome.

WBS

After the primary initiation steps are completed and a scope is developed, now a WBS should be generated. A WBS is a work breakdown structure and includes the deliverables and sub-deliverables of a project broken down into individual work packages. A good rule of thumb is that each work package should be 8 to 80 hours in duration. [4] In this manner they are small enough to predict an accurate count of hours but not every minor task that needs to be done to accomplish the deliverable such as making a phone call, copying files, etc. These are assumed to be understood by those expert in the task they are assigned to. Preferably, the SMEs help determine the work packages because they will then be committed to their estimates.

Essentially the WBS is a laundry list of expected deliverables that make up the entire project. Planning this stage takes experience from a broad-based project team. It should focus on the big picture and as you drill down through each deliverable, specifics will become obvious. For large projects the WBS may need to be revisited throughout the planning phase to assure that steps were not omitted. Should a step be omitted, this will incur additional costs or time and put the project over budget or over schedule. *Fail to plan, plan to fail!* [5]

PND

A precedence network diagram or PND then sequences the work packages in a logical order. Sometimes called a network diagram, a logic diagram, a precedence diagram or a precedence network logic diagram, they are one and the same. One might consider this a logical step, but in the past there have been innumerable project managers that just jump in feet first and start working on various aspects, in many cases the most enjoyable pieces. That leads to a number of issues such as work stoppages, rework, larger budgets or late schedules. These issues can be avoided by taking the time to identify logical timing to prevent mistakes or delays.

Ultimately the PND may be shifted to parallel path a number of tasks, called fast tracking or be used to determine when the most likely completion date will occur. This will be the ultimate schedule baseline. However, individuals are needed to meet the new PND. If you have never performed a PND exercise, here is a brief attempt to demonstrate [6]:

Figure 6.1
Organize the tasks in logical sequencing and with relevant predecessors identified for a housing project.

Figure 6.2
Forward Pass—Earliest Times

How soon can the activity start? (early start—ES)

How soon can the activity finish? (early finish—EF)

How soon can the project finish? (expected time—ET)

Figure 6.3
Backward Pass—Latest Times

How late can the activity start? (late start—LS)

How late can the activity finish? (late finish—LF)

Which activities represent the critical path?

How long can it be delayed? (slack or float—SL)

Figure 6.4
Final PND with combined information for forward/backward pass

If you have never performed an exercise like this, seek out some support on your first attempt. There are some practical benefits from performing exercises to accomplish a feasible PND. Any scheduling software will do much of this work for you, but understanding the root of the programming is similar to learning how to do a derivative or integration in calculus the long way before you learn the short cuts. It is worth the few minutes to learn this properly.

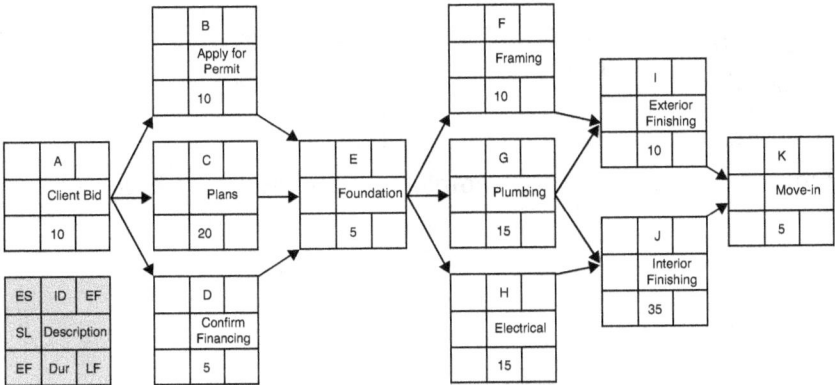

Figure 6.1 Initial Duration and Sequencing completed

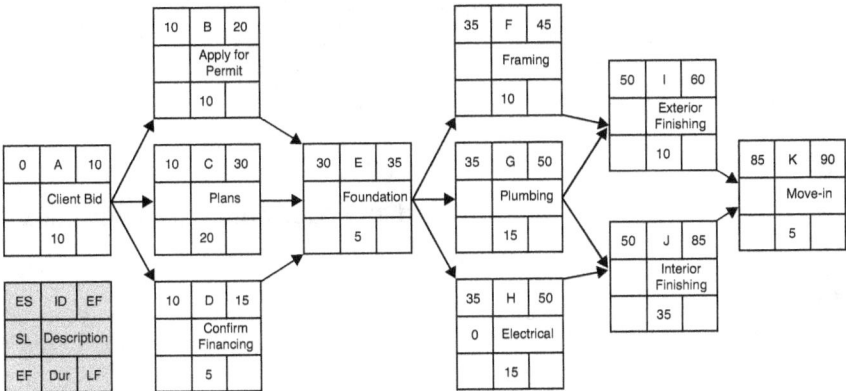

Figure 6.2 Forward pass identifying earliest start and finish dates

Note: this is the "non-additive" method. The PMBOK 6th edition demonstrated a "+1" from the EF of one task to the ES of the next activity. For example, from activities B, C, D to E, the early start of activity E would be 31 (30 + 1). In software, it does this automatically.

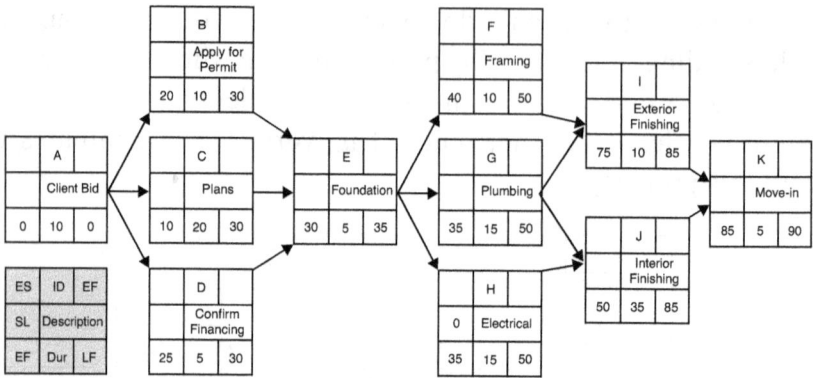

Figure 6.3 Backward pass completed with latest start & completion dates

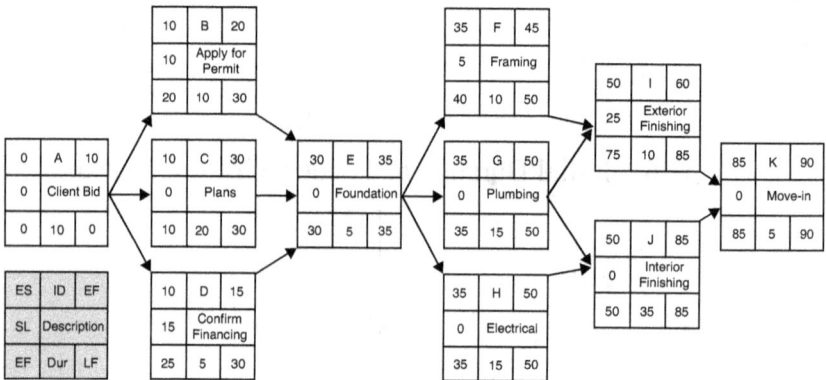

Figure 6.4 Full PND after forward and backward pass with slack

RAM

After the initial PMO meeting, review of past projects, evaluation of stakeholders, scope definition is signed off, WBS and PND are created, the project team needs to divvy up the major deliverables. If there are limited resources available, even if the work can be completed in parallel paths, the work may require sequential ordering to commit available resources. A responsibility assignment matrix or RAM is used to clearly communicate who is responsible for each of the pieces of the project puzzle.

Project management is as much an art as a science and as such requires the human capability element to be practical and successful. Each of the individuals and roles identified below have multiple responsibilities within a project, but we will only address specific responsibilities related to project cost analysis and estimation (Figure 6.5). Organizations may assign different titles to the individuals who complete these tasks, but this is less relevant than the tasks performed by the individuals.

Consider that the RAM chart below only covers the information related to cost estimations, each of the tasks such as scope development, schedule development, quality metrics, reporting, etc could have additional tasks assigned to them. This can also be done with a PMIS that has multiple functions integrated. In any case, the intent of the chart and the placement of the RACI format should help complete this.

There are a number of other project management general processes involved, but the scope of this survival manual will not allow expansion at this juncture. But do spend some time evaluating:

- Formal scheduling procedures
- Crashing
- Resource leveling
- Fast-tracking

Risk Planning (Figure 6.6)

Another area that could be explored further is risk planning, but beyond this minimal amount of information, we will not have time to address this. Here is a basic checklist of the risk planning process:

RAM (RACI Form)	Position Title								
Activity	Project Manager	Project Team Member	Functional Manager	Professional Cost Estimator	Subject Matter Expert (SME)	Prior Project Manager	Coworker	Project Sponsor	Key Client
Seek and include resources	A	R	R	I	I	C	I	R	I
Cost estimation method	A	R	C	C	C	C	I	I	I
Identify strategies for improved estimates	A	I	C	R	R	R	C	C	
Identify prior risks and known pitfalls	C	I	C			A	R	C	C
Identify new risks to cost estimates	A	I	C	C	R	R	C	C	C
Identify cost constraints	R	I						R	A
Review client budget availability	A	I		S				R	R
Document fiscal budgeting needs	A	R	R					I	I
Determine management reserves	C							A	R
Champion funding throughout project	C							A	R
Expert support on specific details	R	C	C	R	A				
Work package cost estimates	R	R	R	R	A	C	C	C	
Review cost estimates of others	A	C	R	R	R	C	C	C	
Identify hidden or duplicate estimates	R	A	C	C	C		S		
Cost estimate summary (living document)	R	A	I	R	R			I	I
Provide feedback to project sponsor	A	C	C	C	C				
Contribute to lessons learned	A	R	R	C	C			C	

Figure 6.5. Responsibility assignment matrix (RAM) in RACIS

R = Responsible A = Accountable C = ConsultI = Inform S = Support Figure 6.5. Responsibility assignment matrix (RAM) in RACIS

Step 1: Risk Identification

- Generate a list of possible risks through brainstorming, problem identification and risk profiling
- Macro risks first, then specific events

Step 2: Risk Assessment

- Scenario analysis
- Risk assessment matrix
- Failure Mode and Effects Analysis (FMEA)
- Probability analysis
- Decision trees, NPV, and PERT
- Semiquantitative scenario analysis

Step 3: Risk Response Development

Mitigating Risk

- Reducing the likelihood an adverse event will occur
- Reducing impact of adverse event Transferring Risk
- Paying a premium to pass the risk to another party Avoiding Risk
- Changing the project plan to eliminate the risk or condition Sharing Risk
- Allocating risk to different parties Retaining Risk
- Making a conscious decision to accept the risk

Step 4: Risk Response Control

Risk control

- Execution of the risk response strategy
- Monitoring of triggering events
- Initiating contingency plans
- Watching for new risks

Establishing a Change Management System

- Monitoring, tracking, and reporting risk
- Fostering an open organization environment
- Repeating risk identification/assessment exercises
- Assigning and documenting responsibility for managing risk

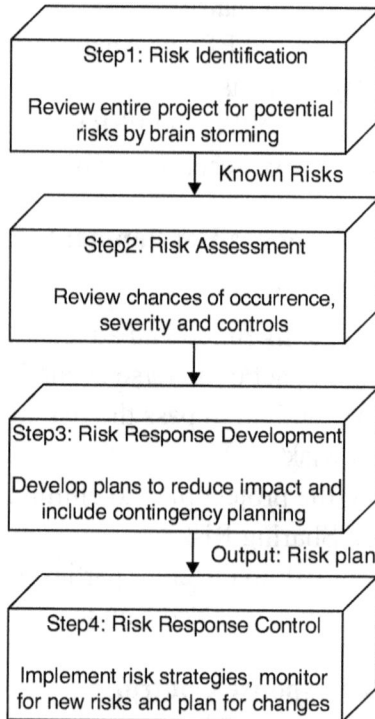

Step1: Risk Identification

Review entire project for potential risks by brain storming

Known Risks

Step2: Risk Assessment

Review chances of occurrence, severity and controls

Step3: Risk Response Development

Develop plans to reduce impact and include contingency planning

Output: Risk plan

Step4: Risk Response Control

Implement risk strategies, monitor for new risks and plan for changes

Figure 6.6. Project Risk Planning Processes

Now we will shift to the specific needs of international projects during the initiation and planning phases. Again, this is not a complete list of everything that a project manager should know for international project management, but in short order, it gives the highlights of many of the crucial steps.

Language Planning

Americans tend to assume that everyone in the world worth working with speaks English or has some comprehension of the language. While the world has been transitioning to English as the preferred business language, it is not universal. To this end, if you have highly capable project team members that are necessary but don't speak English, plans must be made to translate or work within boundaries that work for the team.

However, what frequently happens is that the project manager or engineers may be sent from the US to the target location and they may be the only non-native speaker. In this case, the PM will be translated to the team. During this situation, the team may speak frequently and quickly in their native tongue and the project manager doesn't get everything translated to them in real time. This will require additional effort from the PM to focus and make sure that if a remote conversation is going on that the translator is directed to the most relevant or important in the room.

To overcome the language barrier and time constraints when I was traveling in Europe, I would ask for two translators. One would translate to me what the question was and when I spoke the other translator would pass that information back. In this fashion, there were almost two simultaneous tracks going. Luckily the two translators were able to adapt to this method and my responses and thoughts could work quickly enough for this to work well in a Q&A session. In the break-out sessions, each translator took a different set of groups and focused on them. We accomplished significantly more in this manner.

What becomes more of a challenge is when the project team speaks multiple native languages. Consider the United Nations and the number of individuals that sit on the floor but have translators sitting around the perimeter of the room. This requires a tremendous amount of coordination to get the team performing at optimal level.

Work with a reliable translator. If they are not working out, tactfully seek a second translator and determine if it was the translator or perhaps *you* that need to work on the communication effort.

Time Zone Planning

Another challenge and benefit is working across time zones. When working with a Swedish or German company we were quite aware of the six or seven-hour time difference and recognized that the 8am to 9am hour was the best opportunity for conference calls. But what about Chinese or Thai manufacturers that work from US closing times to a few hours before US opening hours? Somebody must be willing to work off of first shift at some point.

A few strategies that have worked include:

- Inter-company projects usually permit the highest ranking official to dictate the most convenient hour for the team, but it may be biased to their working preferences
- Alternating the after-hours meetings from one company to the other, each meeting, will leave the obligation evenly split and neither party is being oppressed more than the other
- The client gets to work during normal business hours and supplier works in the early morning or late evening
- Two companies negotiate that for six months the one organization will work late hours while the next six months the other company or department will work late hours

Find out what works best for everyone and be willing to adapt. The other benefit is that projects don't go on forever, so even if you are inconvenienced for a period, it will change, eventually.

While the challenges may be meeting one another at a convenient time, one company has learned to leverage these advantages. When a design project is assigned they have offices in Shanghai, Frankfurt and Los Angeles and assign one or two team members from each location to work through the time zones so that the project is being designed 24/5 or even 24/7 when needed. The one designer leaves email notes for the next designer, and fifteen minutes later the project continues right where it last left off. This is a tremendous benefit for the clients of this organization.

One other nuisance occurs if it is neglected. Don't forget that the US time change for daylight savings and return to standard time will occur at different dates than in other countries. For instance, China never changes their time and those in Europe are sometimes a week or two different than the US. Make sure if you have a weekly call with various countries to plan this into the communications plan.

Culture-Training

In the 1980s I had the privilege to work in an environment where we communicated via satellite transmission and view the cultural differences in Japan and USSR. It may seem odd since we had been in conflict with Japan only forty years earlier during WWII and we still were at odds with communist USSR. It was four or five years prior to the end of the Cold War. Yet, we got to share with one another both language, clothing and cultural differences. During this time, to ease the differences, we learned both languages enough for table manners at lunch and to say "Excuse me, I need to use the washroom". We also learned Russian art and Japanese origami. After these experiences, we got into actual cross-cultural education.

If you have the opportunity, make sure that the first meeting is in person and intentionally have both or all represented cultures prepare something that shares their culture. When it is personal, it will help individuals open up and share who they are. This builds the team camaraderie that will be necessary throughout the project and will not be possible on a daily basis.

Another great experience was training an engineer from Hong Kong to become the primary liaison for all manufacturing efforts in Southeast Asia, including China, Japan, Korea, Thailand and Viet Nam. He and an assistant sourcing agent visited the US for two weeks and we had the opportunity to spend virtually 8 to 10 hours each day working on equipment, documentation procedures, troubleshooting and introducing them to all of the US-based players in their company. After hours we took them to local US tourist sites, various restaurants and shopping. This built a relationship that has lasted and is actually quite possibly stronger than many domestic relationships because a genuine bond was built.

In order to assure that each of the individuals in the project team are successful, make sure that they have the tools that were identified in Chapter 3 and 4 by online, in-person or written training materials. The stories related in this book are just the beginning!

Travel Issues

Also recall from prior chapters that travel can be a unique situation and teaching individuals about the roads, foods and other challenges will be interesting. A few other notes, which might seem obvious include:

- Identify whether a passport is needed
- Identify whether a visa in necessary
- Identify the travel routes in advance whether via train, taxi, rental car, flights, etc
- Identify travel safety risks and be sure to adjust for them
- Identify the people you will meet by exchanging photos in advance
- Identify if there will be holiday traffic or seasonal issues (rainy season, etc)
- Identify the exchange rate and amount of money to take, noting inflation as well
- Identify tourist sites you might want to visit if you have down-time
- Identify clothing required by culture or temperature
- Learn the basics of the language such:
- "Hello"
- "Yes"
- "No"
- "Do you speak English?" (or an alternate language that you speak)
- "I don't speak   (language)"
- "Where is the washroom?"
- "Could you write down directions to this address?"
- "Where is the police station?"
- "Does this contain (nuts, dairy, other allergens)?" (if this pertains to you)

There are probably a lot of other phrases that you will pick up in a short time. That will come with the territory and project.

Software Tools

With the significant advancements in Internet, wireless and satellite technologies, one should be investing wherever possible to accelerate the benefit of communications devices, but this should be obvious to most participants. Email, phone conference calls and video conferencing have been popular for twenty years already.

The tools that should be investigated with some diligence are project collaboration and data management. Tools such as Webex can be used and stored during the meeting for later retrieval. However, one also needs to note how project files, technical data and other documentation will be stored, retrieved and protected. Marcus Goncalves wrote a book called the Knowledge Tornado, which was released with an effort to promote the sharing and protection of critical company information. [7] All too often individuals will make a statement that they know that something was done in the past, but finding that information will be too difficult and they will overlook some research that might be beneficial.

Thus, make sure that as a part of the planning effort that individuals are aware of how to store, retrieve, share and protect information so that it is available on demand for the appropriate individuals, but not available to external individuals. There are software packages available for this, but many companies already have an internal process and a PMO should be able to indicate the direction on this. If not, work with the PMO to develop this for future groups.

Such tools will require protection that covers international security protocols which include freedom of exchange, but with encryption and identification tools. Since this is not an IT course, we won't go into any further detail than to point out that "the needed information and only the needed information" should be released.

Agile Tools

In the last ten years there has been a significant movement to integrate agile tools with predictive tools outlined in this chapter. This means that individuals have to learn not only traditional / predictive / waterfall methods, but also the flexible tools from agile.

Agile focuses more on bringing the customer value and embracing change as needed. The plan is developed by a list of features that are desired and will be added to over time. Instead of one long plan, we work through iterations which usually are one to four weeks in length. Every iteration we focus on having a deliverable product. It doesn't mean that the work is actually published, but it could be if we needed to. After each iteration (or sprint) we have a meeting (sprint review or scrum) and evaluate what was completed and which features remain.

There are many different variations of tools that fall under agile including Kanban, Xtreme, scrum of scrums and other tools. There are numerous agile classes and books for this purpose.

Realize that most projects use a combination of predictive and agile tools which is then called "hybrid". Very rarely can any project claim to be exclusively one or the other. The 7th PMBOK coming out in 2021 will focus more the combination of these two methodologies and best practices.

Chapter 7
International PM Execution and Controlling Tools

The first 90% of the software code takes 90% of the development time. The remaining 10% of the code takes up the other 90% of the time. "Cargill's Law" Tom Cargill, Bell Labs [1]

If a project is properly initiated, staffed and planned from scope, schedule, budget, quality and other bodies of knowledge, the project has a good chance of succeeding. However, if a project stops with the plan, then it is merely a good idea and will never produce the end result in a useful product or service. So, how does one efficiently make the plan happen? Through communication.

Every step of the project can be planned, but if the right people, material and equipment are not in the right place at the right time to effectively execute that plan, then the plan will be ineffective.

Over the years of teaching classes on project management, heavily for mechanical and chemical engineers, one issue has to be brought up repeatedly. Engineers that become project managers need to learn when to take off the engineer's hat and put on the project manager's hat. Engineers are interested in making the best product, service, equipment, etc. that will serve the customer. In general, we find ourselves inventing or improving one more function each day. Several years ago, when I was focused in my mechanical and chemical engineering disciplines, somebody put a sticker on my desk that said "Sometimes you just need to shoot the engineer and begin production". It's not a bad thing to desire the perfect product and strive for continuous improvement. However, as a project manager, the focus must be on delivering the work in the scope and *only* the work in the scope on time and on budget. These characteristics are conflicting, but necessary.

Regardless of the conflict, many engineers isolate the two disciplines and only perform as a project manager or only perform as an engineer to keep life simple. Frequently others learn to mingle the benefits of both career paths and transform into some of the most technically-oriented expert project managers available. However, there are specific talents a project manager needs to succeed in engineering projects.

Terminology

The execution elements of project management therefore require a basis in the discipline that you are managing (pure mechanical design, product development, structural engineering, civil engineering, chemical engineering) and the field in which you practice (facilities, commercial products, retail products, industrial products, business-to-business services, event planning, etc). You don't need to be an expert in each area of any given project, but to effectively communicate you need to make sure that individuals understand the nomenclature used. I was once told that "nomenclature doesn't matter; it's only people trying to show off how smart they are." My response is the same to each individual "if you think terminology doesn't matter, go ahead and tell your local policeman that you 'picked up your drugs from the drug dealer this morning' when you had picked up medication from the pharmacist. I bet there is a different response!" Basic knowledge of the discipline and terminology are key.

Communication

As the saying goes "What we have here is a failure to communicate." [2] Besides planning efforts, communication is the next most important aspect of the project. As noted in the introduction to this chapter, if people, materials and equipment are not where they belong when they belong there, delays and cost increases are inevitable. To ensure that people, materials and equipment are where they need to be, communication is an absolute must. This is probably obvious, but what is less apparent is precisely how to communicate. One project manager may heavily favor using email while another may use conference calls only. Yet another project manager may be inclined for a successful project to have as many in-person meetings and require travel to do so.

The right answer for communication mechanisms is usually somewhere in between, but also heavily dependent on the project itself. One may find that a project team located in one area of the world may require a weekly meeting with the heads of each department or group and the project manager can effectively orchestrate these meetings. In between, there might be some emails for a particular individual or group as reminders or information planning. If the team is spread across the globe such as the example with China, Germany and Los Angeles, then a weekly meeting isn't possible and even weekly conference calls would be difficult. In this case a blog may make sense so that all parties can see what has been planned and reply with thoughts.

The other issue, besides determining which method is appropriate, is the necessary communication frequency. One project manager had a weekly conference call regardless of how much work had been accomplished and somehow it miraculously always lasted one hour on the mark each time! When someone recommended to the project manager that we skip the holiday week since there were only two workdays and not much would be accomplished because even those days had few employees on site, he was angered and absolutely would not hear of skipping the call. His call consisted of three or four instead of the usual twelve participants and nothing was really accomplished. He was not using the call as a review of status for the project team members, but rather for himself. He could have spent a few minutes reviewing the documents and gone on with his day rather than using four more hours of company time. Communication should be planned for a "JIT and JAN" approach. This means just-in-time and just-as-needed.

Another error is that individuals are intentionally or accidentally forgotten in the communication. Perhaps a mass email goes out and one or the other is forgotten on the copy list. If they have a critical element to be accomplished in the next time period and they didn't get notice of it, only through luck or the grapevine might they find out. This has happened many times to support functions or consultants that are on the fringes of the project.

On the flip side is the persistent "send all" or "reply all" email individual. Over the years, it is unavoidable to find an employee or even a project manager that will send email or update to every stakeholder. This is a blatant waste of time for the majority of the stakeholders. If the information is highly detailed, it will probably only relate to specific individuals assigned to the project team. If it is a little more generic, perhaps management would like to see and discuss. If it is a milestone update with high-level detail, then perhaps a full stakeholder alert would be suitable.

Take a moment to review the information provided and consider:

- If you've forgotten somebody or if the communication is a waste of time.
- How frequently updates should be made.
- In which method they should receive the information.
- As mentioned, in the planning section, remember to select language use as well.

It will reflect well upon you as a project manager to be considered an excellent communicator. Note that in the US, that a monthly meeting might be suitable, but for other nations you might require a weekly group communication.

Procurement

An additional aspect to maintaining the supply of people, materials and equipment is that related to procuring external resources of each nature. Traditionally the procurement role is not dramatically different for external compared to internal resources, except for the contract. The contract is the binding document, not the scope, which holds all internal employees and resources bound to achieving the necessary work product.

International project management offers the same challenges of domestic sourcing efforts, but with additional challenges. Procurement of staff may require work visas if the company does not have a presence in each country. It may also require establishment of a corporation in that country to operate or receive permits related to the specific work. Some countries sanction foreign business ownership, while others do not. In too many circumstances individuals just get rolling, offer one or the other individual a contract for employment or temporary consulting only to find out that it is prohibited or that middle-men are required to monitor the arrangement.

Taxes become a challenge because both nations or multiple nations may be involved with the product, service or labor rates that are charged and each country is different with the particular tax, tariff or code fee. Another related item is the HST coding of specific international products. If you are interested in HST codes and connecting on this level, check out International Trade Commission for more details. This is a highly specialized area, but usually an import/export company with a good reputation can help select the right codes for your product. [3]

On top of the inbound and outbound taxes, tariffs, port fees and such is the actual transportation fee if you are moving product from one nation to the next. Overseas shipping from Southeast Asia is not cheap. Fuel costs during 2007 and 2008 had skyrocketed to almost $5/gallon for regular unleaded and the cost of diesel for transport matched that. Those that made plans in 2006 for transportation costs at $2/gallon were suddenly hit with tremendous shipping fees and for larger or low margin items this put the squeeze on making a decision whether international production made sense. Also, shipping issues are related to dock strikes, sea conditions and saltwater damage which can make either the product late or damaged. Fuel prices jumped again in the 2010s, but in 2020 due to the COVID-19 outbreak, for the first time in many years, multiple states have gasoline for less than $1 per gallon.

Oh, yes, and more is "bought" in some countries than just the resource. Ethics in contract negotiation is a challenging topic. As already addressed earlier, the ethics of various nations will play heavily. Purchase of some components which are knock-offs, illegal or prohibited by trade agreements also offer challenges that are not as prevalent domestically. [4]

Recommendations would include having an SME or subject matter expert involved in contract negotiation with the countries involved. In this manner it will relieve the project manager of learning the specific terms that each country mandates. You may still wish to familiarize yourself with the contract, but taking on this responsibility yourself may increase project and corporate risk. A reputable negotiator will be well-worth the money to set up a long- term contract.

Schedule Details

International project schedules will be generated in a similar fashion and require deliverables, sub-deliverables and work packages to be identified on the schedule with precise milestones. One item that is important with international project schedules is that the overall schedule will be likely longer due to translation issues, shipping concerns and a plethora of challenges that are not seen domestically. If the schedule must remain intact, then costs will be escalated. Caution would be best in generating the schedule with an understanding that best case, worst case and most likely scenarios should lean toward the worst-case side. This holds true unless significant experience would substantiate that the most likely timeframe will be held with some certainty.

So, what do you do if the project appears to be slipping and the schedule is important? There are two terms in project management that apply whether domestic or international that can be used for accelerating a schedule.

The first method is fast-tracking, which takes sequential tasks and arranges them in parallel, where possible. The benefits to fast-tracking are that it usually does not require any extra funds, since the same work is being performed with the same resources, and that the work package estimates have been well defined in prior planning exercises. However, one of the challenges to fast tracking is the increased risk of rework if the secondary task is partially dependent on the work of the primary task. This means that in the event the primary work package leads to rework, the secondary work package will now also require rework and both will affect the schedule and the budget. An example would be that of building a home. After the foundation and frame are completed, the electrical, plumbing and telephone wiring could all be run at the same time, but if the plumbing bursts and damages all of the work, both need to be redone. One note with fast-tracking is that it may lead to one critical path being shortened, but a new critical path is generated, or even two parallel paths at the same time.

The second method is called crashing. Crashing means that additional resources are added to expedite the work. The benefit is that the additional manpower does not generally add risk because the same tasks and project sequencing. The downside is the absolute additional costs to the project for adding the auxiliary resources. If manual ditch digging is performed, twice as many men and shovels should mean half of the work time. However, if the work is one day long with a backhoe, it would not make much sense to bring in a second backhoe with rental fee and operator to cut the work to one day.

One easy way to separate these two terms in your mind is to think of two cars. Fast-tracking is running two cars in parallel while crashing will certainly cost!

Below are some exercises on fast-tracking and crashing. Should they be confusing or completely new to you, take the time to practice some exercises or complete a practical course in project management before continuing your project management career path. Optimal crash time will keep you from "over-spending to reduce time".

Goals of Crashing

- Find total direct costs for selected project durations
- Find total indirect costs for selected project durations
- Sum direct and indirect costs for these project durations
- Compare additional cost alternatives for benefits

Activity ID	Slope	Max.Crash	Direct Costs Normal Time	Cost	Crash Time	Cost
A	$200	1	5	$500	4	$700
B	$400	2	9	$800	7	$1,600
C	$300	2	10	$600	8	$1,200
D	$250	3	11	$500	8	$1,250
E	$300	1	11	$1,000	10	$1,400
F	$300	2	8	$600	6	1000
G	$0	0	7	$800	7	$700
			Total Direct	$4,800		

Figure 7.1 Crash time costs

Determining Activities to Shorten

- Shorten the activities with the smallest increase in cost per unit of time
- Assumptions:
    The cost relationship is linear
    Normal time assumes low-cost, efficient methods

Crash time represents a limit—the greatest time reduction possible under realistic conditions

Slope represents a constant cost per unit of time

All accelerations must occur within normal and crash times

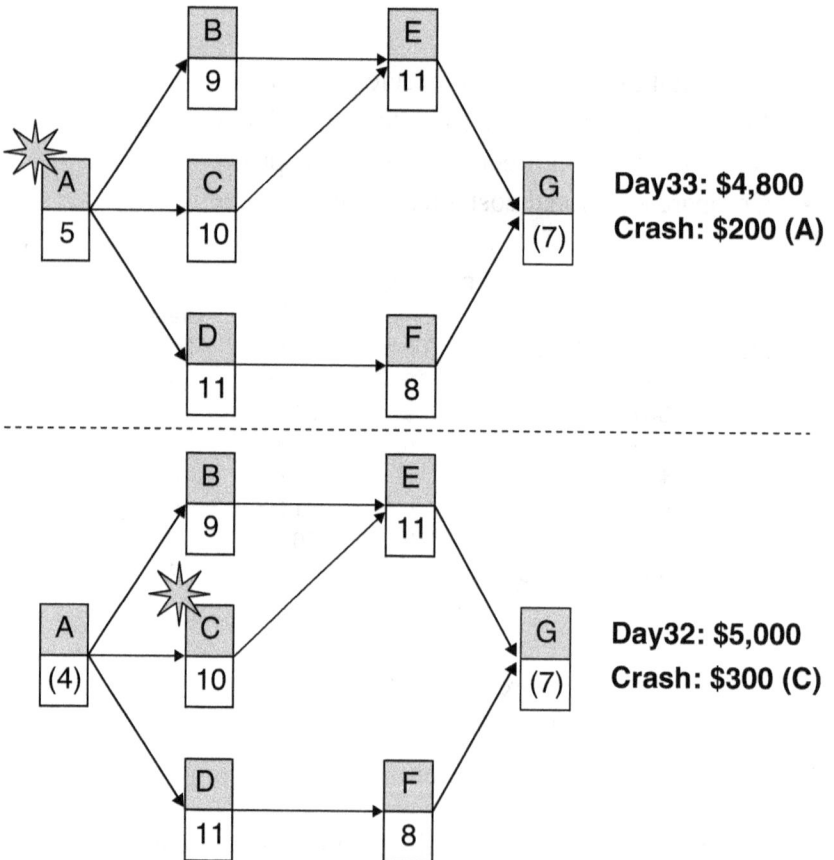

Figure 7.2 Crash Times from 33 to 32

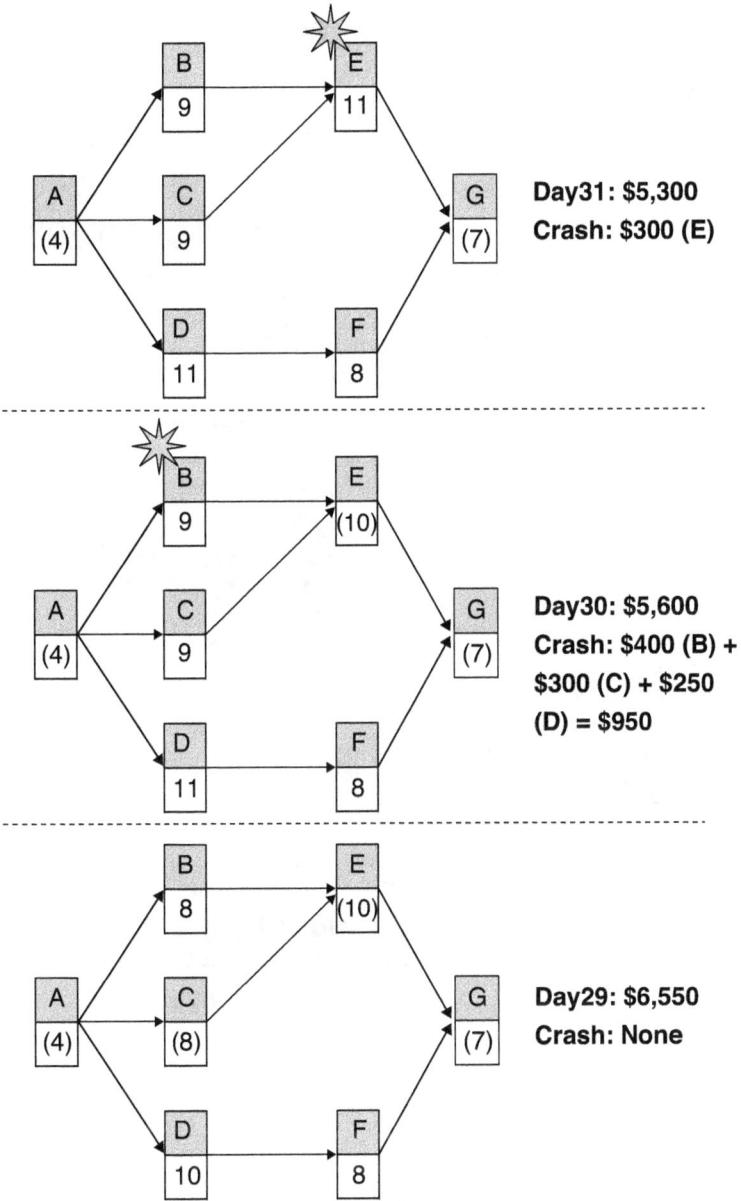

Figure 7.3 Crash Times from 31 to 29

117

Project Length	Direct Cost	Indirect Cost	Total Cost
33	$4,800	6000	$10,800
32	$5,000	5500	$10,500
31	$5,300	5000	$10,300
30	$5,600	4500	$10,100
29	$6,550	4000	$10,550

Figure 7.4 Crash Time Cost Summary

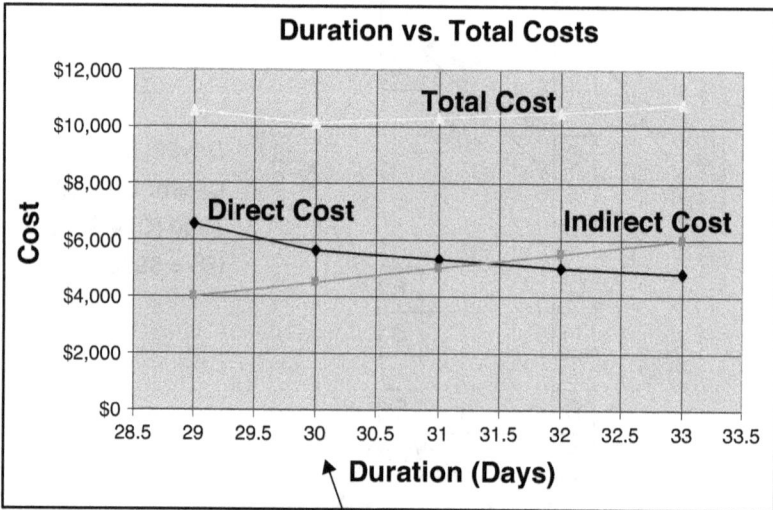

Figure 7.5 Crash Time vs. Cost Chart

Quality Control Tools

So, now that we addressed some of the execution tools and response tools, we also need to address the control mechanisms and response in case a project is falling behind schedule or running over budget.

There should be specific criteria identified when the resolution will be left to the project manager and when the project sponsor or stakeholders should be involved.

Monitoring Frequency

Firstly, the monitoring frequency should be defined for the entire duration of the project. How often will status reports be issued on scope, schedule, budget and quality metrics? How often will risk, communication, HR, procurement and other planning bodies be inspected for status?

All of these items could be left to a 0-100% inspection which means that they are only monitored at the beginning and the end. This is acceptable for short projects or projects with low risk. During the project, none of the inspection steps are considered. The significant advantage is that the costs of controlling are eliminated. Another possibility is to measure at 0, 50 and 100% of the project. At least at the halfway point of the project each of the metrics is checked. If the project is significantly ahead or behind schedule, then steps can be taken to adjust to meet the scope, schedule or budget requirements.

For larger projects, the most frequent method used in project management is "percent complete" which monitors the actual percentage of work accomplished up to the inspection date. If the project is a year in length, perhaps weekly or monthly reporting is necessary. If the project is five years long and only has low-to-moderate risk, it might be permissible to check only monthly or even report each of the criteria on a quarterly basis since the cost to report a project of significant magnitude may cost more than the risk involved.

A basic decision in how frequently to monitor is based on the fact that the cost to inspect project quality should not exceed the risks that it is meant to protect from. If it takes 100 hours work to determine that you are now 20 hours behind, you probably would be better off not checking! If it costs 10 hours inspection to save a 3-month delay on a time-critical project, it is well-worth it! A project manager and the team assembled are expected to make this judgement. Project management domestic or international probably requires the same diligence in this area.

One other item to point out here is that control charts are used to determine when a project manager can resolve situations independently or when others need to be involved. Some individuals use bulls-eye targets that indicate when a project is "out of control" and whether the PM takes steps, the team as a whole takes steps or additional stakeholders are required. Other control charts include linear charts that permit project tasks to continue as long as they are within 5% of targets budget or schedule, or a trend line of 7 periods is not consistently upward or downward trending.

Earned Value Management (EVM)

One other significant area of project control tools would be in EVM or earned value management. This method uses metrics against the plan to compare whether a project is succeeding or failing. This text cannot even begin to explore EVM in great depth, but here are some of the basics and a few charts to help understand what is happening.

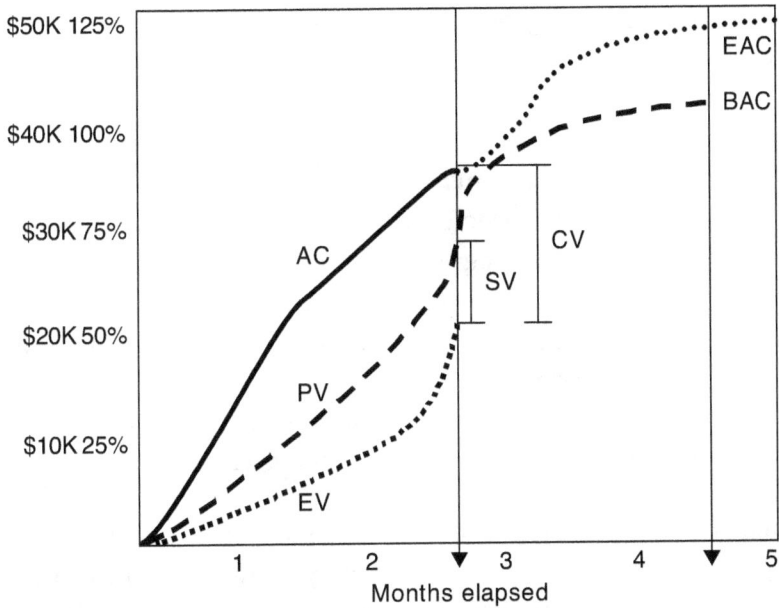

AC: Actual Cost EV: Earned Value PV: Planned Value CV: Cost Variance (BCWP-ACWP) SV: Schedule Variance (BCWP-BCWS) BAC: Budget Cost at Completion EAC: Estimated Cost at Completion FAC: Forecasted Cost at Completion LOE: Level of effort VAC: Variance at Completion (BAC-EAC)

Figure 7.6 EVM Basics

The last piece to ponder is the success rate or yield rate. When one considers various metrics, high yields are not acceptable, but instead, excellent yields are necessary.

If a project achieves 99% of its goal, one might think that this is acceptable, but consider the following statistics, which I've never checked for validity, but illustrate the importance of project management and diligence to hit the scope, schedule and budget precisely.

99% yield would mean:
- 20,000 lost articles of mail each hour
- Unsafe drinking water for almost 15 minutes each day
- 5000 incorrect surgical procedures per week
- 2 aircraft landings that miss the runway at each major airport every day
- 200,000 wrong drug prescriptions each year
- No electricity for almost 7 hours each month

When you consider these statistics, it is scary to consider how some projects miss their target so dramatically. However, it is also easier to understand why projects run over budget and over schedule if they are absolutely required to hit a 100% success on the scope as demonstrated above.

Agile Considerations

Much of the communication and other tools remain the same in execution. The monitoring elements are different as we focus on features being done rapidly rather than on "crashing" type of considerations. There is little concern in agile for tracking to the original plan, because, well, it's not really a fully-fleshed out map but rather a compass. Think about the "tracking" in a traditional plan from New York to Los Angeles with every stop and every turn planned. With iterations used in agile tools you plan as far as you can see. Perhaps you plan the first stretch to the mountains. Then, plan to the next peak and then "re-plan" for the next peak after that. It's more... well... agile!

Lessons Learned

Finally, remember to complete a lessons learned document at the end of a project. While the lessons learned should be recorded throughout the project for each individual, a summary meeting to consolidate those lessons is important. If you wait until after a five year project, many lessons would be lost, so keep it going throughout the life of the project. This should include items from each of the knowledge bodies and process groups. For technical projects we usually identify the product lessons learned separately from the project areas. For example, a wedding would have technical areas such as:

- Venue
- Minister
- Caterer
- Flowers
- Cake
- Band or DJ
- Dress
- Bridesmaid dresses
- Tuxes
- Limo
- Others

All projects should include the BOK:

- Integration
- Scope
- Schedule
- Cost
- Quality
- Risk
- Resources

- Communication
- Procurement
- Stakeholders

The project manager and project team should make a first attempt at the document including as many references to specific items in the project as possible. This would include "when estimating costs for commodities, the project team should have considered the inflation rate of the international country". The goal of the lessons learned document is not to point fingers, but instead to help the team learn from successes and mistakes and keep a single document for reference that other project managers can learn from.

Chapter 8
Individuals Suited to IPM

"Soft skills" are so important for our profession, yet courses in communication and negotiation struggle for enrollment while technical courses immediately fill up. After doing a self-assessment, many Project Managers may realize that they need a course in communications more than one in earned value. (Alex Brown) [1]

The intent of this chapter is to reveal character traits for successful project managers, not to discourage anyone from becoming involved in international projects. Each and every reader should be able to review, at some level, their own personal preparation and find areas upon which they can improve. Project managers from six months experience to forty years' experience should be able to extract useful lessons from these stories.

Another clarification is that there is no perfect project manager. Even within an industry, there is no perfect ideal project manager. Perhaps on one specific project the PM managed to achieve the scope, schedule and budget precisely and the client is absolutely thrilled with the performance attained. This does not guarantee that the next project will run as smoothly or that with a different set of individuals, metrics, or parameters that this project manager will run flawlessly. While one might expect another excellent performance, there might be another individual more suited to the unique challenges. The best PMs will treat every project and situation as though it was unique and requires the special attention it deserves, rather than grow complacent.

In general, though, there are traits a project manager will carry to be effective across a broad industry or multiple disciplines. This list is not necessarily complete or exclusive and any additional thoughts you might have are important for your development and welcome to the author as well. Even those involved in project management for some time will be open to new and different methods of initiation, planning, execution, controlling and closing techniques.

Traits in an Effective Project Manager

✓ *Systems thinker*

Generally, the individual best suited to project management is concerned about getting systems correct, rather than focusing randomly on one detail and the next will be dealt with when it becomes an emergency. There is a process to planning, execution and controlling and therefore after each project they can become more efficient within the systems they oversee.

✓ *Personal integrity*

Individuals without the highest level of integrity will find themselves struggling to achieve the scope of the project. Why? Project managers rely on the project team, stakeholders, SMEs and a number of suppliers to achieve the ultimate goals. If the PM sacrifices integrity, others will see this behavior and become hesitant to support the goals of the project manager without further explanations. Shortcuts and questionable behavior will hamper results. Don't believe it? Just try it, and you will learn the hard way, as many others have!

✓ *Proactive*

As noted many times, planning is a key element of the project management agenda and therefore, those that plan ahead and maintain a proactive approach will be prepared for unusual circumstances. Too often planning is rushed and situations that could have been planned for are overlooked and cost more time and effort in a reactive approach. If you are proactive and run through what-if scenarios, this is a good sign.

✓ *High tolerance for stress*
There it is. Blunt as can be, but there it is. Project manager is a high stress position to hold. If you cannot handle stressful situations or react quickly without considering the ramifications of instructions given under the heat of the moment, you might be best to learn stress management first! The project is in the hands of the project manager. The team has some accountability, but in the end, the project manager is the responsible party for delivering the end-product. Carefully consider the size and complexity before accepting an assignment and you will be better off. International projects add to the stress level by factors, not percentages.

✓ *General business perspective*
The stereotype of engineers does not lend to understanding profitability in an organization. Engineers tend to be black and white or pass/fail in mentality. While this may be a stereotype, most engineers are proud of such logical assertions. However, an engineer involved in or transitioning to project management must understand the ambiguity of business management for profitability and accept a level of "close enough" or a suitable level of detail for making decisions. If not, analysis paralysis will make the individual ineffective as a project manager.

✓ *Good communicator*
Another area identified in prior chapters, but cannot be overlooked here, is communication. If you are shy, avoid sharing information

with others, sit by yourself at lunchtime, or generally prefer isolation, you probably are not best suited to project management. However, if you desire to share ideas with others, work in teams, mentor less experienced workers and learn from everybody you meet, then, communication may be a strong suit for you. This does not intend to indicate that you must strive to be the center of attention. Rather, that you share thoughts and ideas when appropriate.

✓ *Effective time management*

Considering that project time or schedule management is a key element of project management, then only those that are focused on making things happen each month, week, day and even hour will find themselves driven to complete the project on-time. Some individuals are simply focused on perfecting each item no matter how long it takes. This is a good trait for an SME, but not for the project manager who should be summarizing the precise times to complete the "perfect" work packages of others.

✓ *Skillful politician*

The word *politician* often brings negative connotations because of shady agreements that are made in local, state and national politics. However, the majority of politicians are out to serve their constituents to the best of their ability. Negotiation is what politicians and project managers must do to achieve goals. Frequently a project manager must request a specific employee to accomplish the work package at hand, thus having a good relationship and something to bargain with will help. Those who are too shy, too blunt or blind to the benefits of negotiation best stand aside and let someone suited handle this.

✓ *Optimist*

Project management is intimidating at times and considering the risk assessments usually are full of potential tragedies, an optimist's outlook will keep the individual and the team positive. Prepare for the worst and then plan for the best. Years ago, somebody related that a general and his army were entirely surrounded. His second in command commented on how bleak the situation was. The general was far more optimistic, shouting to his men that "it was the best day ever. No matter where they would shoot, they were bound to hit the enemy!" That's optimism.

These are broad traits for all project managers, but what about international project management and all of the material already presented. What makes someone distinguished enough to handle an international effort for their organization and deserve the chance to prove him or herself capable of delivering the "win"? Again, the following traits are not all-inclusive, but rather a good start to self-assessment or assessment of a potential candidate for an international project.

Traits of an Effective International Project Manager

✓ *Work experience with cultures other than one's own*
One solid indicator of a potentially qualified candidate is someone who has already worked successfully with another culture. This may not be related to project management, but those that are married to someone from another culture, have studied overseas in another country, volunteer in a cultural education program or perhaps have supported church missionary trips all demonstrate openness to something new and different.

✓ *Previous overseas travel*
Any experience in travel to foreign nations will help. All of the issues identified with travel in the early chapters of this book will be less difficult for somebody that has seen several countries

before. Even if the project intended is not one of those nations, the individual has learned lessons of how to approach new situations that can be modified for each new country.

✓ *Enjoy travel?*

The big question is whether the individual wants to travel. Some executives have been required to travel, but would much rather not have to leave home, it was simply required for their position. If this is the case, don't seek out international projects. However, for those that love to travel, they will find the experiences in international project management exciting, interesting and fulfilling, not an obligation. My personal disappointment is never having enough personal time on the trips to see the tourist attractions on weekends or in the evenings.

✓ *Good physical and emotional health*

Somebody who is significantly overweight or extremely tall will have a difficult time traveling. Airplane seats are only so big and for those that are tall or heavy, they will arrive far more tired than if they could rest or work comfortably on the flight. Anybody who has several medications might be at risk if they forget a medication and cannot get a replacement in the nation which they are visiting. Somebody with significant food allergies must also question their own suitability for international business travel. Also, those that have emotional challenges will only find this work more stressful. None of these should prohibit anyone from international project management, since some work is done 100% remote, but for those that must travel, consider these thoughts for your own benefit.

✓ *Knowledge of a host nation's language*

Of course, anyone that can speak the host country's language is at an advantage. Business can be conducted more quickly and with less expense by omitting translation. Again, this doesn't preclude non-speakers, but it is a benefit for those that do.

✓ *Recent immigration background or heritage*
From a purely cultural standpoint, anyone who recently left a country and began working in the US, might be considered the optimal candidate to lead a project in their home country. The expectation is that culture shock would not be a concern and all of the issues with language, attire, etiquette and ethics are well understood. A recent immigrant might also have contacts who could be useful for the project as well.

✓ *Ability to adapt and function in the new culture*
For those that are not from the nation, those that show general adaptation characteristics in everyday work would receive some consideration as well. While one does not need to change their own personal beliefs, habits or traditions when visiting another nation, they do need to respect and be willing to meet individuals half-way. This adaptability will help in negotiation and understanding the culture so that planning efforts are done in the right frame of mind.

✓ *Ethnocentric vs. Worldview*
Anybody that has an ethnocentric view that the US is the "best" country in the world and every other country does it wrong cannot function efficiently in international projects. Believe it or not, only a few years ago an individual stated that they had never left the US and didn't see any reason why they should. They already live in the best country in the world. While patriotism is noble in principle, it will only create discord for an international project manager. My approach is to find something of pride for each nation visited and make sure to balance the pros and cons of each culture into the project.

✓ *Attitude*

As mentioned for the general project management section, optimism is necessary, and an overall positive attitude is needed. Challenges will abound from every angle and meeting each of them face-to-face should be exciting. For those that derive personal achievement from overcoming challenges, international project management is a playground. The right project manager will be one who seeks out positive people, positive situations and when everything else is going wrong, takes the smallest of successes and cheers for them! This is not easy to do, but a trait worth seeking in your international project managers.

✓ *Training required*

Even if someone appears to know a significant amount about a country, they would be wise to review specifics prior to a trip. Review language skills, maps and names/faces of individuals to be met. Running through scenarios would also be a benefit. One of the charts [2] used in our training courses is shown below for benefit.

Information-giving Approach (Less than a week in the environment)
- Area briefings
- Cultural briefings
- Films/books
- Use of interpreters
- Survival-level language training

Affective Approach (1 to 4 weeks)
- Culture assimilator training
- Role-playing
- Cases

- Culture shock: Stress reduction training
- Moderate language training

Experiential Approach
- Assessment center
- Field experiences
- Simulations
- Extensive language training

Benefits of International Assignments

Some may ask why project managers desire international assignments. What are the benefits on top of the aforementioned benefits? Let's look at them briefly as well.

✓ Increased income

Generally, if somebody carries many of the positive traits identified, that individual is worth more to the company they represent and therefore, one would expect higher income. Retaining top talent is difficult, particularly in engineering and project management, which remain some of the best career paths for graduates today. Despite many challenges in the economy in late 2008, there is still a shortage of qualified professionals in either discipline. If you carry both a PE and PMP certification, it is even more beneficial to seek international projects, which will open up new levels of income possibility.

✓ Increased responsibilities

Most organizations consider international work the next rung or tier above domestic projects. The new position requires the same management, planning and process experience, but adds new challenges and responsibilities which make the position fulfilling. This may or may not be the circumstance for everyone, but for those that press themselves to achieve new things or frankly just get bored easily, international work can last a lifetime.

✓ Career opportunities

Besides advancement, international work may allow an individual simply to transition from engineering into project management or from one discipline into another. Also, to get a promotion to the next management level in the company, conceivably some amount of experience in international work is required. If you will supervise those performing international work, you should have some level of understanding for the challenges. So, even those that do not wish to remain involved in international projects should have a limited period to broaden their experience.

✓ Foreign travel

Some individuals simply love to travel and project management can take individuals to the most bustling of cities to complete infrastructure projects or perhaps to the most remote villages of India to help battle hunger or provide education through wireless networking. Taking photographs in each location will help make work seem more enjoyable. The labors and efforts are eased simply by knowing that it is all part of a well-fulfilled life.

✓ New lifetime friends

Another wonderful benefit is meeting new people and getting to share culture, not just during a trip, but far beyond this. Some individuals still contact me after many years and we share holiday cards and stories. Possibly this is because both individuals are involved in something special to them, or maybe it is just a bond that is forged because of the time and commitment to the project. In any case, these relationships may also be beneficial years down the road should you be looking for work via employment or consulting.

134

Challenges of International Assignments

✓   Absence from home/family/friends

Beyond the challenges of the project itself (scope, schedule, budget, communication, etc) there is the human aspect to consider. Taking a one or two-week vacation is an excellent experience, but if the assignment is for an extended period, it is quite different. Also, for those that have broader experience, you may be traveling from one time zone to the next continuously for months on end. This will eventually lead to some remorse related to missing family, friends and neighbors during special events and holidays. Unless you are truly the lone horse and thrive on independence, there will be some effort needed to overcome the loneliness of international travel. At one point in my career I spent eight months traveling through the US, Europe and Asia for business. On a few weekends that I was home, time had to be spent transferring laundry, handling bills and getting prepared for the next round of travels. It can be quite tiring and planning for a balance in work and the rest of life will be important.

✓   Personal security risks

As noted in earlier chapters, there is a personal security risk. On a trip to East Africa I was riding in a vehicle in the back seat resting when suddenly the driver began to slow. As I looked ahead a road-block was set up and the driver was pulling to the side. The policeman had intentionally selected our vehicle to pull over. We weren't sure why, but it was apparent that they had chosen us specifically out of traffic. As the policeman stuck his head into the car he looked into the back and said "Hello!" with a bright grin. The gentleman that I was traveling with knew him from prior trips and he just wanted to say hello. My heart stopped pounding, but after this trip I was told that his visit prior and visit following that one were not as lucky. He had been imprisoned on each of the other trips for no specific purpose and let go only a few days later. He was lucky not to have been beaten, despite losing a significant amount of cash both times. All types of stories more or less dramatic happen each day and accepting these risks is part of the role.

✓ Missed career ops

While international projects frequently open opportunities, they also may limit options. Sometimes a multiple-year contract is signed to work on the specific international assignment and when the big promotion is available, you are not permitted to apply because of contractual obligations. Maybe there are no obligations, but you simply are not aware of the positions that were opened and filled during your absence. Therefore, missed opportunities are a possibility as well.

✓ Foreign language issues

Learning a foreign language doesn't happen overnight and it can be difficult to adapt to this even over an extended period. Until you master the language, it will take more effort and tire you more quickly than if you spoke your native tongue. For those that know the language, it still does not guarantee dialect or accent issues will arise.

Depending on the length of stay, the amount of culture shock may vary. Individuals that have been in one culture from childhood on may find the differences too much to handle in one moment. They are not prepared to view, hear, smell and experience everything all in one quick shot. For those that enjoy travel, they may eat it up for some time, but eventually the differences in culture will override the benefits and some amount of adjustment will take place. This cycle is fairly typical, but some individuals adapt in less than a day while others I know personally took over five years to fit into their new environment.

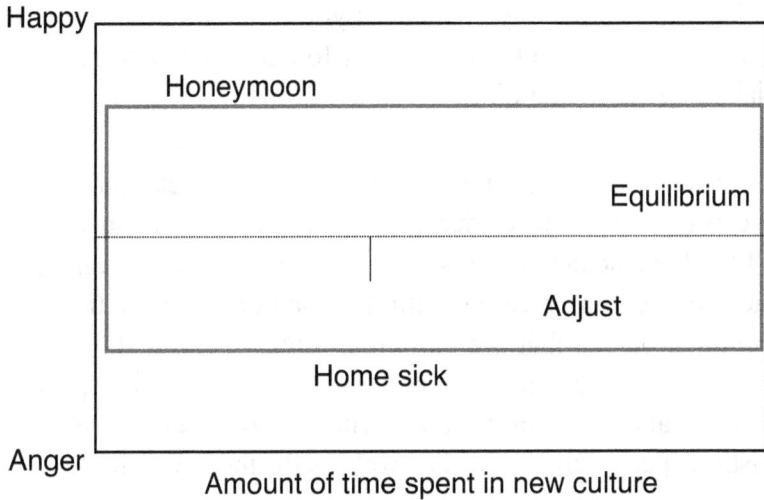

Figure 8.1. Culture Shock Cycle

So what can you do to avoid or reduce the unpleasant effects of culture shock? Here are some thoughts that will help:

✓ Create "stability zones" that closely create home

A majority of individuals do not immediately convert from the existence they are used to in the native country to the new country. To help cope with the dramatic differences at work, travel and shopping, the home is decorated as it was in the native country. It is not unusual for the entire apartment or home to resemble the same decorations, furniture layout or other aesthetic similarity. This helps to make the remoteness seem less dramatic and relieves the internal and mental stresses that the new environment will stimulate.

✓ Modify expectations and behavior

Recognize that you will be expected to behave as the local population does after some time. If you visit for a week, you are not expected to adjust your entire way or life, but you might be expected to learn when to speak, where to sit and how to address people of different social status when at a dinner.

✓ Redefine priorities and develop realistic expectations
Make sure that the priorities established for you as an individual are based on achievable results. Make sure that enough time is dedicated to communicating with those at home to keep the bonds.
For human beings that are extremely family -oriented, emails and calls may occur several times per day. Others may only speak once per week and feel comfortable with the separation. Prepare for realistic expectations for you as well as for those you leave behind, particularly a spouse or children.

✓ Focus on most important tasks and relish small
  accomplishments
At some point you may ask yourself, why am I here? During this time period pay attention to the successes that you see. Even if they are small steps that are moving the project forward, celebrate them! Don't allow work to be a drudgery, but rather celebrate accomplishments and keep track of the positive items.

✓ Use project work as a bridge until adjusted to the new
  environment
This may not be too difficult for an international PM. Usually the demands of the project will ultimately capture your attention and rather than feel lonely, you make "your work your mistress". Fill in empty time that makes you feel uncomfortable by helping another employee, mentoring someone, or learning some new aspect of the project that you did not anticipate engaging during the project. This intentional distraction may help you to get by.

✓ Engage in regular physical exercise programs, practice meditation and relaxation exercises

Besides proper exercise, try to eat a balanced diet, even if the food is different than from "home". Exercise and diet issues will lead to poor mental focus and reduce overall efficiency. Once health is compromised and efficiency drops, stress increases. Eat right and exercise will help more than many individuals may wish to admit.

✓ Keep a journal

In order to avoid the uncomfortable loneliness, write in a journal to keep your thoughts flowing. In many cases, this therapy is all you need to remember why you are doing what you are doing and recommit to the cause.

Take a long hard look in the mirror and evaluate your personality, wants, needs and desires before accepting an international assignment. If it is right though, you will be quite pleased.

For a true international project manager, they will be interested to know that there is now an International Project Management Day which is celebrated on the first Thursday in November. The intent is to spread the growth of the discipline.

https://www.iil.com/international-project-management-day/

Figure 8.2 International Project Management Day logo

A few tools for helping identify fit as an international PM:

Disc Profile:

https://www.discprofile.com/what-is-disc/overview/

FIRO B: Identifies relationship modes

https://www.discoveryourpersonality.com/firo-b.html

Myers-Briggs – indicates your "type"

https://www.myersbriggs.org/my-mbti-personality-type/mbti-basics/home.htm?bhcp=1

Chapter 9
International PM Perspective

A carelessly planned project takes three times longer to complete than expected; a carefully planned project takes only twice as long" Unknown [1]

After having worked with individuals and projects from each continent, well not Antarctica yet, it would seem practical to share some further experiences and perspectives that have helped learn personal growth lessons. At the end of every project the lessons learned is a key document, which will help reduce struggles for each project manager in the company on future projects. Many of these examples are related to relationships more than just the core scope, schedule and budget, but this is frequently where project managers need the most work.

Smart People Learn From Their Mistakes; Wise People Learn From Others' Mistakes

Some of the work that I performed early on in my career was project work. It was not treated this way though because my education was purely in engineering with a minor in chemistry, but no experience in project management. Practical universities today would do well to add a course in project management so students that graduate from engineering have the basics in this practice. Most students will either be on a project team or lead a project team within a short timeframe and the technical aspect is only one element of their responsibilities.

Within fifteen months of graduating with a bachelor degree, the company had made me a supervisor for the engineering group. The only positions younger than me were the interns or cooperative employment students. Some of the older engineers who had twenty years' experience asked "how did you get so smart, so quick"? It's not that I'm any smarter than the next person, but rather willing to learn from everyone around me. When something didn't work right for another engineer in the group, I had taken note. This wasn't to point fingers or to impress my own ego. It was necessary to learn the required lessons and improve my personal performance in future situations.

You can learn from every coworker, supplier, client or stakeholder. This includes the janitor to the CEO. Whether they behaved well or behaved poorly. Whether they behaved too quickly or too slowly. Whether they behaved too emotionally or too mechanically in a specific situation. Observe where success and obstacles impact others and you can avoid the same pain and frustrations they had to go through.

Attitude, not Aptitude Determines Your Altitude

When you perform your daily duties and do so with a cheerful and positive attitude, it will tend to reflect a more cheerful group of individuals around you. For those who choose to complain, see doom at the end of the tunnel and gripe about every situation, they will likely experience similar behaviors around them.

For those that believe conditions are all perception-based and we see the world as we wish, wake up! There are good and bad situations in the world. The behavior is usually the result of a *combination* of the situation's climate, the individual themself and the individuals around them. Perhaps the situation can be seen as a threat to the status quo in the group and many will react with fear.

- We do not have control over the situation.
- We do not have control over how others react.
- We do however have some control over how we react.

If we are conditioned to respond in a certain way by those around us, we only magnify our own traditions. Prior to joining a company, some individuals are incredibly positive and after working only a year in a highly negative group, they have found they are behaving in much the same way as those around them.

The bottom line is that we become creatures of our work environment and if we choose to have a positive attitude to overcome, we have a positive influence on the environment around us.

Every Action has an Equal and Opposite Reaction. Most of The Time.

We may find that project team members are great. Others may seem difficult. Relationships are complicated, but one example that helped me understand mental closeness was related to a physical exercise. In my hands-on training courses, I have the group stand up and perform the following.

1. Find a partner
2. Decide which is the larger partner, height and/or weight
3. The larger partner is to stand with both hands in front of them, flat, toward the other partner. (Assume you are putting them on a wall in front of you)

4. Ask the smaller partner to put their hands against the larger and push
5. Push a little harder
6. Stop

Has anybody ever fallen over? No. In all of the classes, none of the participants that practice this exercise have ever fallen over in fifteen years. Why? As soon as the smaller individual pushes on the larger, the larger begins to push back. Not because they were told to do so, but because it is human nature! They also do not tend to push the smaller person over, despite the fact that they are larger. The force given is equal and opposite!

Another exercise for the class is the following:

1. Have the larger individual make a fist
2. Ask the smaller individual to figure out how to open the fist in 5 seconds

After about 5 seconds of the smaller individual trying to force open the fist with their fingers, ask them to stop. Usually, nobody will accomplish the goal because as soon as the larger person says "ouch" the smaller reduces the amount of force and figures they cannot do it without hurting someone. The one exception was a class in which a larger person nearly broke two of my fingers and despite me saying "ouch" quite loudly, continued to force my hand open! What is interesting is that despite my knowledge of the secret, I continued to hold my fist tight and risk injury. Human tendency is fight or flight. In my situation, I fought.

What is the point of this exercise? After everyone releases their fists, I select the largest and/or most docile individual in the room, approach them and ask them to make a fist. Then I ask someone to time us. As soon as the clock starts I politely smile and ask "would you kindly open your fist?" Sometimes they humorously say "no" and when I say "please" they agree because I asked so nicely! The point is that we sometimes force situations because we think we know better, but asking nicely is far more effective.

Will the approach always work? No, sometimes the reaction is lesser or greater, but usually they are equal and opposite, and nobody is out to hurt the others involved.

Only Burn Bridges You Never Want to Cross Again

In general, don't burn bridges with individuals or businesses because you never know when you will bump into one another in industry again. Oddly enough, in the first ten years of my career I ended up seeing a number of former coworkers that were terminated in other companies, either through contracts or competition. In fact, in one interview for a position as director of a division with 130 people, the president of the company asked if I had worked with a gentleman in the past since we both had the same company on our resume. We had actually worked together on multiple projects and quite frankly did not get along as well as hoped. In any case, the president of the interviewing company asked quite directly what my thoughts were on this gentleman. An uneasy situation for me. My suspicion by his tone, and the fact that the former coworker didn't work with them anymore, was that they didn't get along either and it would be to my advantage to downplay his capabilities so that we were on the same page. However, he also may have been testing to see how negative I might respond. "He was a character." That was my response. It seemed to be an acceptable response because he chuckled and said "That's the way I describe him."

Never did I end up meeting my former coworker, but it always reminds me that when he left, I had intentionally burned the bridge with him because of his behaviors and the peace he took from me and the team. However, it could have cost me a high-ranking position at a reputable company if the president had indeed found him valuable in some capacity. Be cautious that the bridge you burn is something you never want to return over again. When in doubt, leave the bridge alone and let it age or deteriorate in peace!

Domestic First, then International

Engineers new to project management may struggle if their first assignment is an international project, considering the unique challenges identified above. It is advisable to begin with a smaller domestic project that has lower risk. Three particular benefits will arise with this method:

1. You will gain more confidence as you work on a smaller project and the intimidation factor is lesser
2. Should you struggle a bit, the stakeholder frustration won't be as great and you can experiment a bit with the various demands of project management
3. Once you get a grasp of smaller projects you can take on larger and more complex domestic projects and eventually work into international projects with a more fluid approach

There should be no shame in this approach, but some of the most aggressive managers will desire the "big win" before they get their feet wet. One thing to consider is that with longer lives and more years until retirement, we have more time to master various subjects. It may not feel this way, but it will take a career to master project management, so learn as much from each experience as possible and goals will be met in the long run.

Build Relationships, Constantly

Some good advice I received was to work on relationships with employees, managers, department heads and executives well before you need them. Some students have seen this as a conniving effort to politically align with individuals that might benefit your future. It has also been called brown-nosing. This is not the intent! Rather than seek out relationships with only those that might immediately generate political benefit, the advice here is to generate relationships with everyone! All levels of the company can be an important place to have friends. Over the years, I've found that friendships with executives have helped maintain a position with the company, but friendships with entry level engineers, chemists and maintenance staff have helped actually get the job done! When it comes down to crunch-time on a project, no matter how much clout we have with managers and executives, it takes the mental strength and stamina of those that perform the work to make it happen. Janitors and security guards are also friendly and it doesn't take much to smile and put in a little effort to greet them.

You never know when the entry-level engineer this year, might be your assistant PM in a few years or even your boss in a few more years. Build positive relationships with everyone and it will benefit your entire career.

The Golden Rule Falls Short

Many people believe that the golden rule "do unto others as you would have done unto you", used since ancient Greek times by various scholars, is a noble and righteous behavior. While it is indeed

147

a noble intent, it falls short of the insight necessary for others. Where does it say that everybody on your team wants the same thing or things as you do? The fact is all sorts of people desire all sorts of different treatments. The platinum rule identifies that you should "do unto people as they would have done unto them." [2]

Find out what each individual desires and fill that need. Some have the need to be respected, others like to win, some like to by loved and others need to feel right. [3] Find out what turns them on and reward them with the appropriate benefit. Some are not after more money, but rather want constant praise. Others are less interested in praise and need a concrete reward.

Avoid treating the project team as one individual when it comes to reward and find the button for each person.

Be Trustworthy, Be Trusting

Trust is a delicate asset. Gaining trust can be very difficult for those that are skeptical and reserved. They may behave this way because of a situation that occurred early in life and they believe that trust must be earned over a long period of time. On the flip side, trust can be given easily, but also lost just as easily if a minor misstep is made. Once you have gained trust, don't be foolish and lose it!

Trust is maintained through good written communication such as email or text. If the team is remote, it will likely use email or an intranet to communicate and proper frequency will reduce the feeling of isolation. When team members feel alone, they begin to feel that information is being withheld and perhaps the project manager isn't being forthcoming. Generate positive goodwill by sending an uplifting email, even if there is no clear technical reason to do so. Don't waste a lot of time by sending a long email, but a "Hey, how is everyone doing?

I've been busy and hope you are all on track. Looking forward to catching up with everyone at our next meeting. If you have something urgent you can reach me by email or phone at this time for 30 minutes, otherwise I'll assume we are all succeeding at our tasks and will let everything ride smoothly!" During some longer international trips, this kept everyone in touch and usually the respondents asked how the "vacation was going". That levity kept us all united during separation.

Pick up the phone. Your voice might be helpful to the team as well. This is particularly true for those that are still tied to the phone rather than email. Make the effort to draw them in. The phone will take distance away as well.

Nothing is better for sustaining trust than through frequent face-to-face contact. Make sure individuals on the project team can see one another from time to time. That may still require web- conferencing, but make it happen! Even engineers need people, despite the stereotype that would have the rest of the world believe we would rather live alone in our own cubicles and offices.

Slow to Hire, Quick to Fire

Regardless of domestic or international status of your project, one issue that many project managers face is the hiring of consultants or employees. It depends on whether you actually have the responsibility for the employee in question or only have dotted line responsibility.

When hiring individuals, usually a company or project team is in a rush to get started so that precious project time is not lost. However, as a planning phase item, the PM and assembled team need to make sure that the right resources are selected for the long-term. Time spent addressing employee qualifications and personality profile will help reduce conflicts and assure that knowledge and skills are suitable for the position at hand. Planning requires time and procurement and planning mandate attention to this matter.

If an individual demonstrates technical incompetence, then they either require training or need to be replaced as soon as possible. It should not be a matter of offense, but rather judged as a need for a better fit. On the other hand, if the employee at hand is in question because of a challenge with attitude issues, they need to be dealt with swiftly. A negative attitude on the team can penetrate into the psyche of the team and generate negative feelings all around. Instead, warn the individual and give specific terms on which expected changes will be measured. If change is slow or apparently disregarded by the warned employee, replace them immediately. Do not allow a dozen team members to suffer for the sake of a single individual. The team should respect the decision as a benefit to them and an incentive to remain positive.

Maybe means Yes, No means Maybe

One of my favorite international lessons taught to me by my wife. We were at the dinner table in Romania with a friend of her extended family. After finishing the first plate of food the hostess came back to the table and was looking to serve seconds. I wasn't too hungry but figured my best response was a "maybe" response and to think about it for a second and say that "this food is good, but I really should not." Before I could even pause to think, the maybe was interpreted as a "yes" and I had an entire heaping plate of food. My wife told me that sometimes the word maybe is considered a polite "yes" and if you say "no", it means "maybe". If you repeat "no" three times or more then it would be interpreted as an actual "I'm not interested, right now, but maybe I'll be hungry again in 15 minutes!"

It was this lesson that I carried with me. Some cultures take the words at precise value and others expect repetition to feel confident the message was received. We have had some confusion and frustration in our marriage because I need to repeat things three times. She was used to ten times, I was used to once. Three times is a pretty good compromise in my favor!

Could you Repeat that for Me?

When working with individuals that do not speak English as a first language it is wise to understand what "Yes" means.

- American meaning of "Yes": I understand or I agree with your statement
- Asian meaning of "Yes": I hear you or I'm listening

My first major project in China required four other individuals with little to no experience on international projects to communicate with a salesman who had not been in business but two or three years with the US and had two major clients. We had worked with him for about three months when we realized that every time we asked for something during a phone conference, the understanding was not thorough. We would ask if he understood the request and when he answered in the affirmative, we were satisfied to move on. However, the issue was that he didn't understand. His "yes" was not an "I understand", it was a "I hear you speaking to me".

After a few calls of frustration we learned that instead of asking whether he understood, it was better to ask him to repeat the request in his own words. Maybe half the time the actual request was understood. Once he even responded "were you talking to me?" He was the only individual that was on the phone on his side and all conversation was directed between two US participants and him! We were told by a Chinese employee of the company that this wasn't entirely unusual, and we would have to both learn to adapt and create our own group culture so that we understood him, and he understood us.

When in doubt, repeat the question or answers for one another to ensure communication is clear.

Meet my Brother, My Uncle and Three of my Cousins

One project required me to spend about three weeks in Southern California in January. Considering I've lived in Minnesota and Illinois my entire life, this was welcome January weather!

The project entailed diagnosing why more than 50% of a product was failing upon delivery to the customer. A quick evaluation of the first nine machines made it clear that there were two wires reversed. The assemblers had randomly decided to install these wires rather than remain consistent.

Rather than ship the equipment from Southern California to Michigan where it was assembled, the manufacturer decided to hire a team of manual laborers to repair the equipment at a friend's facility in the area. This friend loaned sufficient warehouse space and helped locate a labor supervisor. The supervisor and all laborers were Mexican-born, which is not a surprise in Southern California. By day three, my labor statistics had identified which laborers were performing and which were not. I approached the supervisor and mentioned the statistics I had found and recommended that two of the sixteen laborers be replaced for the next two and a half weeks of work. He asked if he could speak to them and see if their performance would improve. I gave them two more days and stated that if they were not more attentive and reasonably paced that we wouldn't have a choice but to replace them.

What came next surprised me. At the end of Friday the performance had barely changed and I recommended replacement to the supervisor. He said that he noticed the issues too, but despite the fact that he liked them, he would tell his uncle and cousin not to return Monday. Whoa! At first I laughed and asked how many of the others were family members. Seven of the sixteen were family members and the others were friends of the family. Since the manufacturer was paying them and I was representing his customer, it didn't matter to me how they were paid or who was performing the work, but my harsh comments might have been better tempered if I had realized they were family.

While the US may discourage nepotism, other countries find the practice acceptable. Check with local companies or experts to understand this so that you are not surprised. A local manager may expect you to hire their family. Be careful to explain that PMI discourages nepotism, but if you interview the individual and find them to be qualified and hiring family is an acceptable practice, then proceed with caution.

Chapter 10
International PM Resources

"I know that you believe you understand what you think I said, but I'm not sure you realize that what you heard is not what I meant." (Robert McCloskey) [1]

One category that seems relevant is a list of additional recommended resources for those involved in project management. You certainly are not required to read every one of these texts or reference every website while reading the book, but use this as a reference section to come back and review when the time is appropriate. Each of the references is intended to address a separate aspect identified in the prior chapters.

www.pmi.org
You will find the PMI website beneficial for some history on project management, the purpose of the institute and details on how to become certified. Even if you don't plan to become certified, you can join PMI to receive the details of the latest books, brochures, lectures and strategies used in project management. PMI is the global resource for project managers.

A Guide to the Project Management Body of Knowledge LATEST Edition
The Project Management Body of Knowledge is the resource for formal project management and is published by PMI. The details in the guide illustrate the principles of sound project management through text and charts. It is somewhat dry to read, but this is the

official rulebook. The 6th edition is available in 2020 and by 2021 the 7th edition is expected to be released with a number of agile tools included. You should consider the electronic copy if you travel frequently. The hard copy becomes difficult to carry on trips. For about $50 to $70 depending on the source, you have the official book on the basics of project management.

PM Network (ISSN 1040-8754)

This is the official magazine, published bi-monthly, for PMI. It includes relevant resources on brief articles that, even for those of us with experience, will use as a reminder to improve focus in certain disciplines. The magazine is current and keeps up with the latest trends and features some of the most interesting projects around the globe. It also demonstrates the variety of projects that are covered from buildings and IT to product development and service projects.

The Discipline of Market Leaders by Michael Treacy and Fred Wiersema (1995)

This book is more than a decade old now, but the wisdom within is still the guiding drivers of modern marketing. Be the best service, be the best value or be the best product in the market. Focus on that and you will win. The examples are becoming outdated since Airborne Express was purchased by DHL, but the points made are beneficial to a project manager. Understanding the marketing value of your project will help you deliver the scope with greater accuracy.

✓   Olympics websites

IOC: www.olympic.org

Beijing: https://www.olympic.org/beijing-2008

Vancouver: www.vancouver2010.com

London: www.london2012.com

Sochi: sochi2014.org

Rio: https://www.olympic.org/rio-2016

Peongchang: https://www.olympic.org/pyeongchang-2018

Tokyo: https://www.olympic.org/tokyo-2020

Why did I include these? I enjoy the Olympics! Also, because of the magnitude of the event and the truly broadest international scope possible. The Olympic Games are an enormous event and there is no "redo". The challenge in managing these events is that they don't allow for rework, they are time-constrained for sure. The balance usually is a trade-off between cost and scope. The 2008 Beijing Summer Olympic Games demonstrated that with virtually limitless funding ($42B USD) that the scope could be enormous, decorative and nearly perfect. The 2012 London Summer Olympic committee had already promised that the games cannot manage to match the majesty of the Beijing Games in magnitude (only $17B planned) but will aim for character. Space in London is also much tighter and thus planning was critical. These will open your eyes to the challenge of a truly international event! Rio offered challenges for the first games in South America. Of course, Tokyo was the first "delayed" games in history. Others were cancelled, but these will be a challenge.

www.asme.org

This site is referenced for the engineering aspect. Those that are reading this text might be interested to know that it is published by ASME Press and there are many other technical and project oriented books in the ASME main site. Resources abound and knowing the benefits of this website will help expand both your engineering and project management career.

www.rosettastone.com
www.duolingo.com

Here are a few sites that can help with learning languages, even if it is a brief trip. Rosetta Stone is "the" site to get the highest quality language lessons, but it does come at a price. Fortune 500 companies and those dedicated to working long-term in a specific country for more than one project need to consider purchasing this series or at least a 30-minute CD or MP3 to learn the basics for the first trip. The second site is a free site with sponsored links. The material is less thorough and I cannot vouch for the validity of the translations and dialects, but it is free and could get you started for a first trip without investing in CDs or other materials. Duolingo online or in app form has been very popular over the last several years. It allows you to do very short lessons at your own pace.

www.timeanddate.com

This site is good to use to check on the time for each of the locations where you are going (even if it is just so that your spouse or special someone knows the time difference!) The real value is identifying national holidays and planning around down time. The challenges mount when you have multiple countries that all celebrate different dates off. Note: you can find out when you are exactly 10,000 days old, international dialing codes and numerous other items that revolve around the time and date globally.

www.xe.com
www.oanda.com
www.x-rates.com

Refer to these websites for current exchange rates, which will be helpful for planning trips or for determining negotiating rates on contracts. Depreciation planning would also be important if there will be a significant amount of time or chance for change in the rate during negotiations. Any of the sites are reasonable, but depending on your preference you should seek out one of them. While www.xe.com is the most popular, I've used www.x- rates.com the most frequently over the last few years.

Financial Times or www.FT.com

This is considered the "World's newspaper" and should be read for practical and current business stories. Take the time to check out the headlines and relevant features. It is a London based paper with a truly global view of financial markets, investment deals and general world news mixed in. Each of the stories is a mature and less dramatized version of the morning news!

www.lonelyplanet.com
www.world66.com

All of these sites will be good for searching out food, transportation, tourism and travel plans. A lot of the basic statistics can be found for the country on the entry page, which lends to understanding just a bit more of the country. Of course, specific sites dedicated for every country in the world cannot be referenced here, but they could be linked from these sites and more details found at each location. As a PM, certainly you will find this information on your own, but these are a few good sites.

travel.state.gov

Before you go, review the WHTI regulations as they change. The WHTI is Western Hemisphere Travel Initiative. This will allow various additional documents approved to be used as evidence of citizenship.

www.USEmbassy.gov

For security, visa and passport details, this is the official site of the US State Department and should be reviewed prior to a foreign departure. After planning to visit Tanzania in early 1999, plans had to be reevaluated. In late 1998 the embassy in Dar es Salaam was bombed and extra precautions were advised by the US government. When abroad, know the address of the embassy and carry a copy in your passport as well, just in case.

www.adherents.com

This site gives a thorough review of major world religions and can be used to gain an elementary understanding of the principles and expectations of the people that live in a culture with influential religious beliefs. The statistics might identify religious beliefs that influence the culture and might identify behaviors you should retain or from which to refrain. There might be other sites such as www.religioustolerance.com or others that influence your view as well, but they tend to be more subjective in stance.

www.international-business-center.com

This site is a non-profit organization involved in educating business managers, and therefore project managers, in achieving successful interaction with their domestic and foreign business partners abroad. It covers meetings, gift-giving, etiquette and many of the topics aforementioned. Specifics can be tricky, but this is a place to start your search.

HELP! Was That a Career Limiting Move? Pamela Holland and Marjorie Brody (2005)

This is a quick read self help book that explores etiquette related to email, phone, meetings, dinners, professional rapport, interviews and use of appropriate humor. This can be read in about an hour or two, but is sometimes comical and other times quite striking when I realize some of the poor habits that creep into my own behavior. Make this a fun read for a flight some time since it is small and enjoyable and doesn't require a lot of focus to absorb the intent and content.

Management Skills for Everyday Life – The Practical Coach by Paula J. Caproni (2005)

This is a more academic text, which will take some focus, but is enjoyable as well. The author delves into the inter-workings of professional relationships and how different personalities interact and sometimes generate conflict. Ms. Caproni describes power distance and some of the behaviors that cause American project managers or managers in general to misstep and refocus based on the international framework within which they are working. Read this in pieces and take time to digest the topics so that you are able to incorporate them into your practices.

NationMaster.com

This site might get you addicted if you have the tendency toward statistics. I didn't know that the average British citizen consumes 42 times as much tea as an Itatian! At least not until I began using this website. This site has been used for the initial project evaluations and understanding the differences from one country to another. One individual or the other might challenge that they already know enough about each country in which they will work, but ask your self this: does everyone on the team have that knowledge? If not, evaluate each of the countries and provide the details to team members. This is a good assignment for a junior member. Send them to this site and ask them to assemble relevant details similar to the data provided for Brazil, China, and India. They may find many other interesting facts that will raise awareness for the entire team.

Just be careful and know when to stop searching. You could spend hours learning that in Romania one case allowed for a 38,584-year conviction to prison! Not useful to my project work, but certainly interesting!

www.brazil.gov.br
www.china.org.cn
www.india.gov.in

These are the official sites of the countries identified in chapter 5. They are not as flowery as some of the additional tourism sites that you will find online. Nevertheless, they are a substantial and trustworthy site to locate official information. Frequently you can link to other more tourism-based sites that will help find fun spots or useful tools in scheduling flights, hotels and transportation.

www.microsoft.com/project
www.primavera.com

www.Monday.com

www.clarizen.com

The most popular scheduling tools can be found at these sites. Microsoft Project is still the most popular of tools for schedule management. It remains the most cost-effective tool and considering the world-wide coverage of Microsoft products, it usually does not interfere with the Windows platforms or other software in the suite. You can effectively set up a Gantt Chart, PND or other format to view the schedule and resource usage.

Primavera or Clarizen are premium packages with greater tools and flexibility which allows the project manager to gain better view into the project performance and useful metrics. These packages are premium and therefore cost more. If your organization is serious about project management and implementing successful tools, it will want these packages or other premium software on the market.

This is not intended to promote any of these software packages or downplay others that exist, but rather to note the popular tools available.

Budget Controls

Every company should have its own function that controls how the budgets are maintained. For many the "budget" is purely departmental, but if you take a full project management course and pay attention to the WBS and estimating sections, you will learn how the departmental budgets (OBS) work into the project through the WBS. There are formal inventory and project tracking software such as Quickbooks, Oracle or SAP and similar packages, but most project managers will need to work within the framework of pre-established financial controls.

In one client organization the financial controls and software were new. To protect the budget means the project team kept everything in hard copy and in an Excel spreadsheet to control the budget as a backup. Luckily they did this because the entire system had glitches and the only way to rectify and clean up the entire project was to use the Excel spreadsheet and hard copies to re-populate the new software. After a few weeks everything was in the system, but it wasn't a pretty process. If your organization doesn't have formal software to track project budgets, only department budgets, then make sure to keep your own records, even if this means a profit and loss type statement in Excel. This is better than nothing.

If you have a full Project Management Information System (PMIS), then the budget materials will likely be covered there. If it is a mature system, then you should be fine. If it is new, consider a backup plan for the project.

PMO Information

Check out the link to some fairly general papers on best practices related to PMOs on the web: http://whitepapers.zdnet.com/search. aspx?kw=Project+Manageme nt+Best+Practices. What you will find here are general reviews of project management tools and techniques, but some are specific to PMOs. If your company has more than two project managers, then it is time to begin considering a plan to begin work with a PMO.

Risk Planning

Project risk planning is a very specific task for each company and project, but if you don't have experience, a place to start would be http://www.method123.com/risk-management-plan.php. Use the templates, which for $9.95 are a bargain. However, realize that over time that you will need to improve these risk templates. These are risks that will affect the project and not necessarily the product itself.

For equipment projects one of the more popular product risk templates is based on the Big 3 American automakers who developed the dFMEA or pFMEA processes. This text and numerous other guides are available through Amazon or local bookstores that will walk someone through their first dFMEA or pFMEA.

PE Magazine offers risks and strategies for addressing construction needs:

Top 5 Risks and Strategies

HIGHEST FREQUENCY RISKS	HIGHEST IMPACTING RISKS	FAVORITE STRATEGIES, TACTICS, AND PROCESSES
1. Commodity demand	1. Estimating accuracy	1. Integrate risk into contract
2. Energy prices	2. Government regulations	2. Use standardization
3. Skilled craftsmen	3. Commodity demand	3. Hire internal staff
4. Estimating accuracy	4. Construction firms	4. Increase meeting frequency
5. Const. service demand	5. Const. service demand	5. Request budget increase

Figure 10.1. Top risks and responses [2]

For engineers, both the project and the product are significant from a risk standpoint. Making the risks clear and frequently re-visited will help reduce the shock factor to stakeholders when they occur.

Contract Support

When it comes to contract management, I'm no attorney, but some of the advice offered earlier would include seeking an expert in the contract negotiation. Since every country is different and within countries the local regulations are different yet, you would be wise to search for an individual or firm with the precise needs on your own. However, make sure that the individual or firm can:

- Demonstrate prior work in that nation
- Require references for work in that nation
- Determine if they will defend contracts in court or simply write contracts for fees
- Review their prior record and ask for win/lose ratio in contractual court cases
- Determine fees in advance, hourly fees can get out of hand, so look for fixed fee options

A reputable firm may already have cut sheets for the country in question or at least be willing to provide this on demand. Their concern should be to demonstrate to you as the client an excellent background in the negotiations but also competence in your particular discipline.

EVM Sites

www.earnedvaluemanagement.com is a site that will show the absolute basics, some templates and then sell you a simple package to perform this work. If your firm has an EVM process, they should be able to demonstrate in an hour or less the principles of the method. It might take a little more time for some individuals to interpret the graphs and data, but the basic concept should be easy enough to follow.

Chapter 11
Good, Bad and Ugly Projects

"Two people can dig a lot faster than one. Dig." (Angel Eyes from The Good the Bad and the Ugly) [1]

Many times, just reading about terms and concepts doesn't make the ideas come to life. We get a partial comprehension, but not the full or clear picture we desire. Humans learn by watching things that work or things that don't. A story or example demonstrated in a successful or failure mode can work a lot better to sink in. In order for you to see the good, the bad and the ugly, here are some examples and resources to read more about projects that went well or poorly.

Panama Canal Expansion

Back in 1914 the Panama Canal was considered one of the greatest engineering feats in modern history. The ability to connect the Gulf of Mexico and the Atlantic with the Pacific took a roughly 13,000 mile voyage from Alaska to Maine down to roughly 5,000 miles. The costs saved were enormous and from both project and engineering standpoint, it was a marvel.

In 2016, after about 10 years of negotiations and work, the expansion of the Panama Canal was completed which significantly increases traffic through the canal. The technology was of course not the same as the original project, but instead has concrete designed for 100-year life and the steel gates for at least fifty years. The project required multiple companies, forty thousand workers and a budget over $3B dollars. This is certainly considered a mega-project.

The six year construction timeframe was completed as planned and roughly 14,000 additional transits per year can be made.

https://www.intechopen.com/books/case-study-of-innovative-projects-successful-real-cases/the-new-panama-canal

One World Trade Center

Considering the facts around the events of September 11, 2001, the completion of the One World Trade Center in 2013 was a victory for pride of the nation. It was a definitive "comeback" New York to let the world know that we will not live in fear.

The building itself is not the tallest in the world, but rather stands a representative 1776 feet tall. The project was of particular difficulty considering that the work initially required so much effort to remove the twisted and tangled mess from the prior WTC buildings and the surrounding structures that were destroyed.

Much of the project work began under ground and thus the public may not have been aware of the efforts until aboveground construction began.

From a purely project standpoint, this was not 100% of a success. It was a scope success and a triumph of spirit, but the costs were roughly 33% greater than originally anticipated and there were delays in the schedule several times. Perhaps Mega-projects are the hardest to find success because they have so many stakeholders and in many cases, one or the other stakeholder will halt things just to demonstrate their power. Not easy.

https://global.ctbuh.org/resources/papers/download/24-case-study-one-world-trade-center.pdf

http://www.cif.org/awards/2015/01_-_One_World_Trade_Center.pdf

Safety-Kleen System, Inc. – Field Evaluations

From my personal experience going back more than 15 years ago, I can share a project of great pride. Executive management had signed a contract for 10,000 units per year in the first couple years and then even more (15K/year) in years three, four and five. The equipment design had not been evaluated nor and testing been performed within the company. During the first year there had been so many mechanical issues and quality problems that the engineering team was struggling to keep up with the issues. It really was a nightmare scenario.

During the second year of manufacturing and already 17,000 machines having been received, there was a new issue. On a Friday afternoon there was a call that the field was experiencing an unusually high rate of leaks from completely new units. This meant that a flammable liquid was being seeped or "drained" into the customers facility and sometimes reached the drain. Terribly concerning as this was not just a reliability issue, but a potential hazardous chemical situation. That Friday call got the emergency alarms going. At 2pm, several of the department heads had a phone call, during which, they conferenced me in as a consultant. I was very familiar as they'd had me work on this

169

equipment before, with the other failures. It also meant that the manufacturer knew me well.

During the call, the request was made that I take a team and visit 600 or more of the machines in the field so that statistical evidence could be gathered of how many machines were experiencing minor or major leaking. Minor was determine to be a leak that was contained inside of the machine; major leaks would be those that exited the machine at all. They wanted 600 machines inspected within two weeks. My first request was for support individuals as I was a single-person consultancy. They denied that and told me to hire staff and get ten people in the field by Monday. After a brief pause, I told them "Give me until Tuesday to have a team in the field". By 6:00pm, eight other individuals were already contracted for the next two weeks. It was a scramble for me to find anyone in my network that could do mechanical work and had two weeks available! At 6:30pm I received a call from someone at church letting me know that they had been let go that day. They weren't pleased when I gleefully expressed "Really, that's great!" After explaining my situation, they agreed to work with me for two weeks as well. The "ten" were complete.

Saturday was spent planning all of the required details and training for the other nine individuals. During the morning, the training materials were drafted. The afternoon was spent buying tools for the team. The evening was spent finalizing travel arrangements for everyone to leave Monday afternoon.

We all gathered at the local branch and I reviewed, trained, practiced and drilled individuals on their assignment. By lunch time all were packed and heading to the airport. Each individual reported to me nightly (after I had visited customers myself) and we entered locations, quantity inspected and results for each site. The days were long, but charged by the hour!

The details of the exact failures may be a bit fuzzy after this much time, but about half were leaking at delivery. The team stayed under the initial budget and were able to visit about 660 units during the scheduled visits around the nation.

Ultimately it was on time, on budget and delivered the results in the format requested by the team. Based seeing so many units in the field, the company asked me to visit the manufacturer and we reviewed the quality process and fixes for both installed and in-house machines, which was going to be addressed within 60 days.

Failed Projects
✓   London City Garden Bridge

Nice ideas don't mean a successful project is at hand! Ideas are just that... concepts that need to be fleshed out with a vision, proper business case and charter. Then followed up by a team of well-established subject matter experts who will clarify the details with a proper scope, schedule and budget to make sure expectations are within the realm of reality.

This didn't happen with the proposed Garden Bridge in London. Back in March 2012 there was a vision to add more character to London and perhaps break up the monotony of stone and brick with building after building and bridge after bridge. The concept to add more character and greenery was "nice". Especially considering the Olympic Games were to take place that summer, all sorts of construction and improvements had been evaluated.

Five years later in 2017 the project was killed before any execution work on the project had taken place. The project was evaluated by the new mayor's committee and found to have "shaky" financial justification and not a clear vision of why the project would even exist. That doesn't bode well for a new project proposal. Without a clear vision or purpose, it was killed and the Garden Bridge of London has not happened.

What can we take away from this lesson? Perhaps the number one element for all project managers: make sure that the business case and charter are clear before digging deeper into planning. Without such, projects are doomed to fail.

Read more about it here: http://calleam.com/WTPF/?p=8985

New Coke

In the 1980s Coca-Cola was looking to make a splash with something new and different. What resulted has been considered one of the greatest marketing blunders of all time.

Considering Coca-Cola has been one of the top brands by value and has penetrated almost every nation on earth, it is surprising that they also carry such a "huge mistake" on their resume as well. In April of 1985, the company halted production of traditional Coca-Cola and launched "New Coke". The idea was to "sunset" the old formula which was considered a great product, but perhaps could be greater.

The company did not make such a blunder just by flipping a switch or on a whim. The company had done taste testing with 200,000 individuals and the New Coke flavor was preferred. Turns out, that wasn't enough! In fact, there was such a revolt from customers and such a drop in sales that less than three months later the CEO announced that "we have heard you" and that the "Classic" Coca-Cola would be returning. This cost them millions of dollars in revenue and could have significantly turned the tides and given Pepsi and RC room to take their market share.

There is a clear message with this case as well. It's not about scope, schedule or budget though. It's all about hearing the stakeholders. This project was initiated because the company thought they knew what was best and would help the customers figure it out.

Any project manager with some experience knows that taking the stakeholder feedback seriously will be a significant key to success or failure. Make sure to listen first, evaluate second, decide third!

Consider too that Coke did most of the research domestically and the response was domestic. How might an international launch have worked out? Could have been even more painful!

Read more about the New Coke failure here: http://wk.ixueshu.com/file/95f9573fbcadea49318947a18e7f9386.html

Challenger (1986) and Columbia (2003) Space Shuttles

One of the most significant news stories of my lifetime was in 1986. Since a child, the space program fascinated me and one teacher in high school suggested I enter the space program and research initiatives. That passion is perhaps why this is hard to share or understand. The Challenger Space Shuttle was taking off from the launch pad and a bit more than a minute after takeoff, exploded with devastating effects to ship and crew. Thousands of individuals were involved in the preparation of this the "STS-51L" project. NASA had become accustomed to the launches and there were always little glitches and little adjustments. In fact, NASA would delay a launch when needed. So how did it come on that cold January morning to be such a glaring failure? Similar to another NASA failure we'll share next.

In 2003, I actually got up in the morning to watch the Columbia Space Shuttle land. After 27 successful missions it was on "Project" STS-107. The crew had successfully completed the majority of the work for the project, but then disintegrated during re-entry killing all seven astronauts aboard. What happened that could have caused a "routine" trip to destroy the ship.

Besides the horror of the lives lost, there were technical and project lessons that we must take away. The engineers will typically first look for root cause failures of technical nature. You can read more on case studies that analyze the technical failures about o-rings failing or tiles knocked off of the Challenger and foam strikes which made an extremely large hole in the leading edge of the Columbia wing. However, there were project failures as well. Consider that the temperature had never been so cold for a launch as January 28, 1986. Some of the engineers suggested a delay of launch, but the PM (mission chief) decided to launch anyhow. During 2003, Linda Hamm decided that even if there was a hole in the leading edge of the Columbia wing, there was nothing they could do about it in space (no such thing as Space Duct Tape strong enough to hold it together!).

Many individuals who have done case studies on these topics have noted that both of these situations were project managers over-ruling team members who were subject matter experts and it came down to team culture. We need to learn from this and be open when a team member disagrees. Perhaps make sure that their arguments are either opinion or statistically well-founded.

Database from PMI

Beyond these few examples provided, the project management institute has a variety of studies that have been completed over time and might be of interest for a variety of areas including government, capital and various other project types. Consider looking for those that relate closely to what you do now for more input on how to proceed.

https://www.pmi.org/business-solutions/case-studies

Chapter 12
Roadmap for the "New PM"

"If you have no roadmap, you have to create your own." (Jacqueline Woodson) [1]

One last chapter that seems relevant is aroadmap to transition into project management. Of course, everyone may take a different path, but we figured that having some plan is better than none!

Image courtesy Nicole V. Porter

Read!

The best place to start on your road trip to a project management career is to read and learn about it. Of course, this book should be a solid starter position. My hope is that it gives a foundation for the basics, but also for international flavors of what to expect when leaving the US.

In general, if you become a member of the Project Management Institute, you should begin using PMI.org. There is a broad array of projects that usually focus on construction and infrastructure, but also on IT, government and events.

If you take just a few minutes, there are so many different places you can find more information about projects in your discipline. If you're in construction, then find online articles and magazines about projects that were completed or in process for small, medium, large or mega-projects building schools, offices, malls and such. If you focus on wedding planning, then read information for local or national bridal magazines. Hopefully you will see both good and poorly run projects that require attention no matter what your type of project is.

Of course, you could also read information directly from the PMBOK that will help you gain the foundations. To be honest though, it's not terribly exciting and sometimes having a navigator is easier to follow.

Classes

As noted, sometimes having a leader who can walk you through the important elements, explain terminology and put the concepts into context is highly beneficial. Items that will be important to you as the student will be the duration of the class, experience of the instructor and the application or amount of hands on work the class will do in teams.

For the past 15 years I've been teaching the PM courses with ASME and enjoy bringing the principles of formal project management to life. The goal is to share the few variety of stories, many of which were brought out in this book. The variety helps each individual to think outside of their microcosm. Projects encompass more than just software or more than just product development.

If the ASME courses were not of interest, or maybe you took those courses and wish to go further, you can pursue online courses at a local college or university. An eight-week or 15-week college course will give you more depth of understanding and help you to know if this is something you *really* want to make your profession.

For me, it wasn't enough to take a class, I wanted a degree in project management to really note to potential hiring parties that I want a career in this area. My choice was Boston University because it was in the top five for project management in the *world*. For me, it was a great choice and helped me gain my next position, officially as a director of two functions, overseeing all projects.

Practice, Practice, Practice

One area that so many individual miss is that they don't have experience and transitioning to a new career can be a risky setback. You may have to take a significant cut in pay. Or you may have to work your way up from the bottom again. That's the price that many have taken.

One option is to volunteer. Find ways that you can reach out to your community and come up with options where you work for free, but helps the common cause. There might be a local charity that could really use someone to coordinate a fundraiser. You may not be a salesperson, but the coordination of the phone individuals is a requirement for a PM. The coordination of budgets and the schedule, which is not always the "caring person's" forte, could be your way to invest. It could be for your local church and helping plan a Sunday School outing. It could be with the local fire department and running a carnival. Perhaps the local hospital could use a "coordinator of volunteers". You would be the volunteer of volunteers! These are just a few ideas that come to mind right off. The mistakes you make will be overlooked, because, let's be honest, it's free labor! Just get out there and start helping to manage SOME project.

Once you've done that, publish something about the experience. Whether it is in the local newspaper to help the charity gain recognition for the next event or not, make sure it's published. Share about the project's scope, schedule, budget, resources, risks and how they were all handled. Even if you are not a champion writer, have someone help to edit. Later on, you can share the story with a potential employer and demonstrate that you were a PM and did so to help people. It's a double-win! If you can, get it published by formal magazines based on the cause. It's a good start.

Certification

Once you've learned, practiced and have some time into project management, then you can consider your certification. As noted in the earlier chapters, there are various levels, most reading this book will either apply for the CAPM (Certified Associate in Project

Management) or the PMP (Project Management Professional). Pay attention because the process requires you to:

1. Complete a certain number of hours in the discipline
2. Fill out the application
3. Pay the application fee
4. Take the exam
5. Start on continuing education
6. Renew every three years

Check for the latest requirements, which do change every few years, regarding the hours and application itself. Also, remember the exam is tested to the latest PMBOK, so know which version and don't be fooled by a cheaper copy online. It may be out of date.

New Career

Finally, you've made it to the point where you can justifiably be called a "professional" project manager. Congratulations!

At this point you will start your efforts toward achieving that new role. Perhaps you can do so within your current organization. Hopefully they will recognize the path to certification or an advanced degree were not easy. In my case, it was all of them (PMP, degree, publications, experience, etc).

If your current organization does not have a position in project management for you to advance or you simply want a different experience or market, then you'll need to pursue your network to advertise your interest. More people get jobs from their network today than from blind applications online. Both can work, but know what market and types of projects that you want to manage. Some wish to keep to their traditional experience. For me, I still dabble with engineering, patents and such, but also have done a lot of events which are exciting to me.

Choose well and choose wisely for a happy road ahead. Remember, even if you made a wrong turn, you can still back up and head down another path to make new experiences!

Final Thoughts

We've now covered the basics of project management, the details of some countries and how to evaluate the locations for benefits and risks. We have explored the initiation, planning, execution and controlling tools in international projects. We looked at lessons learned to conclude or "close" projects. Finally, we looked at individuals, stories and tools that will fit into an engineer's path to making international projects work.

Just as engineering is an on-going learning process for an entire career, so too is project management. Honing skills to work with the technical aspects of the project may come naturally but the communication and personality skills usually require a little extra work. Make the effort and it will pay off.

An interesting article was in the Mechanical Engineering magazine related to how the engineers of the Apollo 13 ground crew came together and solved several problems in a short time-frame. This example was at the advent of modern project management. The closing arguments of the event note "If we have veterans to thank for our freedoms, and firefighters and police officers to thank for our personal safety, then we must thank engineers for our standard of living." [1] To this I will add, "engineers that can deliver their product on-time, on-budget and within the scope"!

Keep using lessons from personal and professional life to support each other and make learning "life-long"

About the Author

Brian E. Porter, PE, PMP, MSPM, a native of Minnesota, USA is the Vice-President of Strategic Partnerships for MGCG. He is also the President of EEE Consulting, LLC based in Fort Myers, FL. Mr. Porter has over 20 years' experience in project management, product development, engineering, safety listings, patent, business strategy and start-up management in computer sales, consumer products, hazardous waste industry, industrial manufacturing and retail product markets.

Mr. Porter holds a Bachelor of Science in Chemical Engineering from the University of Illinois at Chicago. He also holds a Masters of Science in Management with Specialty in Project Management from Boston University. Today, he teaches at Boston University, Nichols College, University of Illinois at Chicago and Elmhurst College in a variety of business and management topics.

Mr. Porter has maintained his professional engineering license in the state of Illinois for several years. He also has multiple patents within the USA and internationally.

Brian is a member of the Project Management Institute and has credentials as a Project Management Professional (PMP). His international efforts include working with firms in Canada, Mexico, Romania, China, UAE, Thailand, Malaysia, Australia, Japan, Sweden, Israel, Great Britain, Egypt, Italy, Germany and many others!.

Mr. Porter has instructed courses for professionals from Siemens, BASF Chemical Company, Hovensa, SBM-IMODCO, Columbian

Chemicals Company, US Coast Guard, US CIA, Disney, Ashland Chemical, PPL Montana and several others.

Brian can be contacted at brianp@mgcgusa.com or brianEporter@ EEEconsultingINC.com

Bibliography

Chapter 1

1. Easynet, Retrieved December 18, 2008 from http://easyweb. easynet.co.uk/~iany/consultancy/proverbs.htm# mgmt
2. Gray, Clifford F., and Larson, E. Project Management: The Managerial Process, 4th Ed. McGraw-Hill: NewYork (2008), p5.
3. Stevenson, William J. Operations Management, 9th ed. McGraw-Hill: New York (2007), p126.
4. Goncalves, Marcus and Brian E. Porter, Global Management Strategies for Sales, Design, Manufacturing and Operations: New York (2008).
5. Kent, Simon. Passing the Torch, PM Network, Newtown Square, PA, December 2008, p37.
6. Allitt, Patrick, The History of the United States, 2nd Edition, The Teaching Company, Chantilly, VA, 2003, part 5 and 6 of 7.
7. Allitt, Patrick, The History of the United States, 2nd Edition, The Teaching Company, Chantilly, VA, 2003, part 5 and 6 of 7.
8. Henry Ford, Samuel Crowther (1922). My Life and Work. Doubleday. p. 72.
9. Edison Inventions, Retrieved October 22, 2008 from http:// inventors.about.com/library/inventors/bledison.htm
10. Westinghouse Inventions, Retrieved October 22, 2008 from http://inventors.about.com/library/inventors/blwestinghouse. ht m and many others:
11. Edison's genius, Retrieved October 22, 2008 from http:// edison.rutgers.edu/inventions.htm
12. Rayport-Jaworski, Introduction to ECommerce, McGraw-Hill: New York 2008, p29.
13. http://www.pmitoday-digital.com/pmitoday/ december_2018?pg=4#pg4https://www.pmi.org/certifications

A Guide to the Project Management Body of Knowledge, Third Edition. Project Management Institute, Newtown, PA, 2004.

Chapter 2
1. Easynet, Retrieved December 18, 2008 from http://easyweb. easynet.co.uk/~iany/consultancy/proverbs.htm# mgmt
2. Aesop's Fables
3. Treacy, Michael and Fred Wiersema, The Discipline of Market Leaders. Addison-Wesley Publishing, Reading, MA, 1995
4. Gray, Clifford F., and Larson, E. Project Management: The Managerial Process, 4th Ed. McGraw-Hill: NewYork (2008).
5. Stevenson, William J. Operations Management, 9th ed. McGraw-Hill: New York (2007), p133.
6. A Guide to the Project Management Body of Knowledge, Third Edition. Project Management Institute, Newtown, PA, 2004.
7. A Brief History of Cyberspace, Retrieved January 10, 2009 from http://www.smartcomputing.com/editorial/article. asp?article=a rticles/archive/l0503/02nb/02nb.asp&guid=
8. www.jurassicpark.com
9. Water Cube Wears its Coat, Retrieved December 7, 2008 from http://en.beijing2008.cn/46/39/WaterCube.shtml
10. PM Network, October 2008 and http://en.wikipedia.org/ wiki/2008_Summer_Olympics
11. Retrieved from http://www.ustr.gov/Trade_Sectors/Services/ Section_Index.ht ml

Chapter 3
1. Easynet, Retrieved December 18, 2008 from http://easyweb. easynet.co.uk/~iany/consultancy/proverbs.htm# mgmt
2. Abroad Spectrum, PM Network, January 2009, p67

3. Speaking a Different Language, PM Network, August 2008, p49
4. Eurotrip, 2004
5. Columbia Crime Statistics Retrieved January 7, 2009 from http://www.nationmaster.com/red/country/co-colombia/cri-crime&all=1Nationmaster-Columbia
6. More Wealth Without Risk, Charles J. Givens. Pocketbooks, New York, 1995, p50.
7. Retrieved December 27, 2008 from (http://www.cnn.com/2008/WORLD/africa/11/19/somalia.pirat es.boomtown.ap/)
8. Retrieved 1/10/09 from http://news.bbc.co.uk/2/hi/middle_east/2936772.stm
9. What price a Chinese Emperor? Retrieved January 10, 2009 http://www.atimes.com/atimes/China/IE11Ad01.html

Chapter 4
1. Easynet, Retrieved December 18, 2008 from http://easyweb.easynet.co.uk/~iany/consultancy/proverbs.htm# mgmt
2. Business Attire, Retrieved January 10, 2009 from http://foster.washington.edu/gbc/businessdress.shtml
3. Wyatt, David. *Thailand: A Short History* (2nd edition). Yale University Press, 2003. ISBN 0-300-08475-7
4. The Borderless Leader, Leadership in Project Management 2008, p11.
5. NSPE Code of Ethics for Engineers. Retrieved January 10, 2009 from http://www.nspe.org/Ethics/CodeofEthics/index.htmlPE Code
6. PMI Code of Ethics. Retrieved January 10, 2009 from http://www.pmi.org/PDF/ap_pmicodeofethics.pdf
7. Management Skills for Everyday Life – The Practical Coach by Paula J. Caproni (2005)
8. Abroad Spectrum, PM Network, January 2009, p65.
9. Help! Was That a Career Limiting Move? Pamela J. Holland and Marjorie Brody, Career Skills Press, Wichita, KS, 2005.

10. Gray, Clifford F., and Larson, E. Project Management: The Managerial Process, 4th Ed. McGraw-Hill: New York (2008).
11. Phillips, Joseph, PMP Project Management Professional Study Guide, McGraw-Hill, NY, 2004.

Chapter 5

1. The Ten Strangest Places to Drive in the World, Retrieved January 10, 2009 from http://www.argusrentals.com/Europe/Spain/Top-Ten-Strangest- Places-To-Drive-in-the-World-(3). html
2. Management Skills for Everyday Life – The Practical Coach by Paula J. Caproni (2005)
3. US Bankruptcy Courts Statistics. Retrieved January 10, 2009 from http://www.uscourts.gov/Press_Releases/2008/bankrupt_newst at_f2table_jun2008.xlsBankruptcy stats
4. The Myth of the Melting Pot, Immigrants Shunning Idea of Assimilation. May 25th, 1998. http://www.washingtonpost.com/wp- srv/national/longterm/meltingpot/meltingpot.htm
5. F. Kluckhohn and F. L. Strodtbeck, Variations in Value Orientations (Evanston, IL: Row, Peterson, 1961).
6. Stevenson, William J. Operations Management, 9th ed. McGraw-Hill: New York (2007)

Chapter 6

1. Easynet, Retrieved December 18, 2008 from http://easyweb.easynet.co.uk/~iany/consultancy/proverbs.htm# mgmt
2. Attributed to Albert Einstein
3. A Guide to the Project Management Body of Knowledge, Third Edition. Project Management Institute, Newtown, PA, 2004.
4. Phillips, Joseph, PMP Project Management Professional Study Guide, McGraw-Hill, NY, 2004.

5. Military proverb, origin unknown
6. Gray, Clifford F., and Larson, E. Project Management: The Managerial Process, 4th Ed. McGraw-Hill: NewYork (2008).
7. Goncalves, Marcus, The Knowledge Tornado, Blackhall Publishing, 2002, p10

Chapter 7
1. Easynet, Retrieved December 18, 2008 from http://easyweb. easynet.co.uk/~iany/consultancy/proverbs.htm# mgmt
2. Failure to Communicate
3. USFTC Harmonized Tariff Codes Retrieved January 12, 2009 from http://www.usitc.gov/tata/hts/bychapter/index.htm
4. Knockoffs
5. Mendlinger, Sam. Boston University MG742 Operations Management, lecture materials, Fall II 2008 course.

Chapter 8
1. Retrieved January 2, 2009 from http://www.alexsbrown.com/ pmpersskilltypes.html
2. ASME/MGCG slides for PD467 Project Management for Technical Professionals
3. Gray, Clifford F., and Larson, E. Project Management: The Managerial Process, 4th Ed. McGraw-Hill: NewYork (2008), p508.

Chapter 9
1. Retrieved January 2, 2009 from http://www.alexsbrown.com/ pmpersskilltypes.html
2. Management Skills for Everyday Life – The Practical Coach by Paula J. Caproni (2005)
3. Gilliam, Joe, How to Handle Difficult People (Lecture), Rockhurst University Continuing Education Center, 2003.

Chapter 10
1. Easynet, Retrieved December 18, 2008 from http://easyweb. easynet.co.uk/~iany/consultancy/proverbs.htm# mgmt
2. PE Magazine, Owners Need to Plan for the Unexpected, December 2008, p19

Chapter 11
1. *Two people can dig a lot faster than one. Dig." (Angel Eyes from The Good the Bad and the Ugly, 1966).*

Chapter 12
1. *"If you have no roadmap, you have to create your own." Jacqueline Woodson Quotes. BrainyQuote.com, BrainyMedia Inc, 2020. https://www.brainyquote.com/quotes/jacqueline_ woodson_882869.*

Final Thoughts
1. Kerno, Steven, The Efficient Frontier, Mechanical Engineering (magazine), March 2008, pp32-35.

www.ingramcontent.com/pod-product-compliance
Lightning Source LLC
Chambersburg PA
CBHW070715220326
41598CB00024BA/3165